Good Seeing

Best Practices for Sustainable Operations at the Air Force Maui Optical and Supercomputing Site

Lisa Ruth Rand, Dave Baiocchi

Prepared for the United States Air Force
Approved for public release; distribution unlimited

For more information on this publication, visit www.rand.org/t/RR1097

Library of Congress Cataloging-in-Publication Data is available for this publication.

ISBN: 978-0-8330-9170-3

Published by the RAND Corporation, Santa Monica, Calif.

Preface

The Air Force Maui Optical and Supercomputing Site (AMOS) is located on top of Haleakala on the island of Maui, Hawaii. The site, which is owned and operated by the Air Force Research Laboratory (AFRL), has three large (meter-class) optical telescopes that were designed to identify and track ballistic missile tests and orbiting human-made objects. Together, these telescopes provide the Air Force with a unique capability to support space situational awareness (SSA) missions in the Pacific region.

Since AFRL took over the site in 2001, AMOS has received the majority of its funding from congressional add-on funds. However, this source of funding is no longer available because of the austere fiscal environment within the U.S. government. To help with future planning, the AFRL Directed Energy Directorate asked RAND to help develop a strategy for making AMOS self-sustainable over the coming years.

As part of this effort, RAND conducted research to gather a set of best operating practices from within the civil astronomy community and the U.S. military. This document presents conclusions from that research. RAND developed a framework to identify the most appropriate analog facilities and researchers, and then conducted interviews with people working at those institutions to learn more about them. Finally, RAND compiled the results into a set of best practices and recommended strategies for AMOS. The content of this document was researched before the completion of a business plan developed by RAND, which has since been implemented at AMOS.

This research was sponsored by Dr. David Hardy, Director, AFRL Directed Energy Directorate. The study was performed within the Force Modernization and Employment Program of RAND Project AIR FORCE. It should be of interest to those working to develop a long-term sustainability strategy for AMOS. It should also interest researchers and policymakers addressing operational research problems associated with running a large astronomical observatory.

RAND Project AIR FORCE

RAND Project AIR FORCE (PAF), a division of the RAND Corporation, is the U.S. Air Force's federally funded research and development center for studies and analyses. PAF provides the Air Force with independent analyses of policy alternatives affecting the development, employment, combat readiness, and support of current and future air, space, and cyber forces. Research is conducted in four programs: Force Modernization and Employment; Manpower, Personnel, and Training; Resource Management; and Strategy and Doctrine. The research reported here was prepared under contract FA7014-06-C-0001.

Additional information about PAF is available on our website:
http://www.rand.org/paf/

Contents

Table

Summary

The Air Force Maui Optical and Supercomputing Site (AMOS), located at the summit of Haleakala on the Hawaiian island of Maui, is a major site of U.S. space surveillance activity. Operated by the Air Force Research Laboratory (AFRL), AMOS is the only Department of Defense (DoD) site capable of providing visible and infrared spectrum images of space objects passing over the Pacific Ocean—a crucial capability for U.S. space situational awareness (SSA). The three optical telescopes operated at the site each have distinctive capabilities that contribute to the overall functionality of the site. The 1.6-meter telescope is used to conduct nonresolved photometry and can be used during the daytime. The 3.6-meter telescope is used for high-resolution imaging and is the largest optical aperture within DoD. The 1.2-meter telescope, not currently in use, can be used for wide-field searches—surveying large areas of the sky and imaging many objects of interest within a single exposure. Once back in service, as AFRL is planning, the wide-field capability of the 1.2-meter telescope will broaden the overall technical capabilities of the site, now limited to the narrow-field capabilities of the 1.6- and 3.6-meter telescopes.

AMOS's mission includes both space observing operations and research and development. For more than a decade, the majority of funding used to run AMOS had been drawn from congressional add-on funds. In fiscal year 2013, however, the funding profile shifted dramatically, with the majority of the operating resources coming from the Air Force along with modest support from external customers. A new funding arrangement began in fiscal year 2014 driven by the Air Force's desire to lower its total operating costs. Under a memorandum of understanding (MOU) between the Air Force Materiel Command (AFMC) and Air Force Space Command (AFSPC), each organization will provide a set amount of funding each year over a five-year period to support the upkeep of the facilities, equipment, and data collection and delivery at AMOS.

In anticipation of this new funding arrangement, AFRL asked RAND to conduct an analysis of AMOS in two phases. In the first phase, completed in June 2012, RAND assessed AMOS's proposed modernization plan and evaluated the outcomes from previous studies of AMOS. RAND found that the modernization plan aligned with customer needs. Phase II of the study will build on the previous phase to develop a value proposition and business plan intended to help the site achieve self-sustainability. The analysis is grounded in a detailed examination of AMOS's historical context, customer base, product lines, and potential best practices. This report addresses the best practices component of the research, intended to provide suggestions for how AFRL might further streamline operations and minimize costs once the initial business plan has taken effect.

RAND has identified a number of specific ways in which AFRL might operate AMOS more efficiently once the business plan has been implemented. These recommendations are distilled from best practices implemented at research institutions that have attributes in common with AMOS. Collectively, the institutions evaluated had the following characteristics: (1) observatories and other sites of research and data collection located in Hawaii, (2) government facilities in remote places that might provide insight on staffing and logistical matters, and (3) research facilities that conduct blended operations and research and development. Based on interviews conducted at 15 sites, RAND identified five best practices most relevant to AMOS's future sustainability and success. We describe each of these findings in more detail below; the findings are also summarized for at-a-glance reference in Table S.1.

Table S.1. Summary of Key Findings

Key Idea	Enabling Factors (Exemplars)
Wide and narrow field of view missions are staffed and resourced very differently	• Specialized staffing (Pan-STARRS) • In-situ engineering (Pan-STARRS)
Enabling remote observation offers an opportunity to lower overhead	• Control over remote sites (Keck, NASA IRTF, LCOGT, ALMA) • Incentives for summit workers (Keck) • Shared support services (Keck, ATST) • Flexible scheduling methods (NASA IRTF, Gemini North, TMT, LCOGT, UKIRT, Kwajalein, ALMA)
Investing in messaging and outreach can bolster financial and cultural support	• Staff engagement and education (MIT/LL, NOAO) • Service/funding partnerships (SWPC, Kwajalein, NOAO) • Public Relations (Keck, NOAO, STScI)
Maintaining a flexible observatory can raise the value to customers	• Configuration management system (Keck, NASA IRTF, TMT) • Hardware flexibility (Subaru, NASA IRTF, TMT, MIT/LL)
Developing data-sharing systems can lead to greater impact	• Data storage and accessibility (Pan-STARRS, UKIRT, STScI)

1. Wide-field and narrow-field mounts are run differently. These differences can affect staffing, time management for operations, engineering, maintenance, and the ways in which the collected data can be used.

AFRL intends to provide the ability to conduct both wide-area searches and narrow-field inspection and characterization, managing both types of telescopes at one site. The differences between these two types of mounts are of particular interest. Wide-field observations are extremely rich in data that multiple users can exploit, which in turn increases the knowledge that can be derived from the data collected by a single telescope. Data taken during wide-field surveys provides *in situ* feedback on how well the instrumentation operates—another advantage of wide-field observing. Facilities that operate telescopes used for wide-field observing benefit from having professional surveyors on staff—technical experts in the art and science of sky surveys—to operate the telescopes, rather than the generalist astronomers who typically supervise the operations of narrow-field telescopes.

AFRL hopes to bring the 1.2-meter telescope back into use to conduct wide-field searches. This type of observing is an entirely different mission from that of the two narrow-field telescopes that are currently active. AMOS should, therefore, be mindful of the differences between the two methods. In particular, AFRL should be aware that scheduling at the 1.2-meter will be different from current scheduling practices in place at the narrow-field telescopes. Less time will be needed for engineering and calibration as a result of the *in situ* feedback, which should be used to help inform calibration and engineering needs. AFRL should also consider employing "professional surveyors" to run the 1.2-meter rather than switching staff regularly between the different telescope mounts on the site. A degree of efficiency may be gained by recognizing these differences and their impact on operations. Finally, data collected by the 1.2-meter should be preserved for later use by potential customers, regardless of the initial data needs of the original principal investigator (PI); doing so can increase a site's value proposition.

2. Some facilities have lowered overhead by expanding remote observation.

We noted the expanding practice of remote observation at several key observatories. But while strong remote capabilities may reduce the cost of staffing and infrastructure needed to support visiting researchers (lodging, transportation to and from the summit, food, and other basic needs), there is also evidence that these cost savings may be transferred elsewhere. For example, sites that require hands-on access to the instruments may have to increase on-site staff to support researchers observing from remote locations. Remote operations can be conducted in different ways—from control rooms at the base of the summit or by researchers in the continental United States. These two approaches require different technical setups and staff.

Remote operations require installation of several footprint systems. In addition to reliable instrument switching capabilities, a steady stream of telemetry must be supplied by detectors and actuators situated at all points of possible failure; remote staff must have constant access to this telemetry. The telescope must be able to be opened and closed remotely, and weather changes must be readily measurable from afar. In addition, a reliable set of operating standards and protocols must be established and strictly enforced.

Despite the best practices observed, remote operations are not yet foolproof, nor is cost reduction guaranteed. AFRL needs to evaluate whether or not to remote the 1.2- and 1.6-meter telescopes based on the following two-step method. (It is not practical at this time to remote the 3.6-meter telescope due to limitations in running adaptive optics remotely.) First, AFRL should assess what technical changes are necessary to remote the site (such as improved scheduling software or instrument swapping technology) and estimate the cost of implementation as well as the eventual payoff in savings. Second, AFRL should keep in mind that expanding remote operations may not result in savings in labor costs—rather, it could instead require investment in different kinds of labor. Using this method, AFRL should be able to determine whether there is an advantage to pursuing expanded remote operations.

3. Greater investment in messaging and outreach can lead to wider internal and external support—both financial and cultural.

Dedicating staff and resources to outreach can result in greater support both internally and externally—both of which are important for maintaining a successful and sustainable research and development facility. In institutions with robust financial support, comprehensive internal messaging has been an integral part of maintaining steady funding. This is because all staff members are able to accurately discuss the technical advantages of the site and how they can benefit potential customers. The Hubble Space Telescope is an exemplar of outreach and external messaging, with dedicated staff publicizing Hubble's accomplishments through various media. The payoff is evident in the level of public awareness of the telescope. Even on a more scaled-down level, effective external messaging can have a positive impact on the consistency of support among stakeholders in a variety of scientific institutions.

Given the importance of strong internal and external messaging, AFRL should consider cultivating a dedicated messaging initiative so that AMOS's unique contribution to U.S. SSA is broadly and thoroughly known inside and outside the Air Force. A comprehensive outreach program would contain two types of messages: (1) general messages sent out several times a year to inform a broad audience of upgrades, promote research produced using AMOS data, and provide detail on notable successes; and (2) individualized messages targeted to the interests of specific personnel engaged in the SSA effort. In addition, AFRL should develop a program to educate its staff on AMOS's mission and value proposition and instill a cultural focus on mission awareness.

4. Maintaining a flexible configuration can raise the value proposition of an observatory but requires robust scheduling capability and a change of mindset.

Many observatories view the fielding instruments brought by visiting researchers as a nuisance, due to the downtime required for setup and the money to maintain them after the researcher departs. But this equipment can add value by expanding the capabilities of the facilities, which in turn means that more types of research can be conducted on the site—a particular benefit to research and development facilities. Thus, allowing flexible configuration, and viewing it as a positive attribute, can add significantly to an observatory's value proposition by expanding its potential customer base as well as potentially expanding the scope of SSA missions.

While AMOS welcomes visiting experimenters and their instruments, the value of AMOS's research and development capability would be strengthened by developing dedicated procedures to facilitate and even court visiting researchers. Yet, realizing the value of flexible configuration requires a change in mindset. AMOS should adopt an approach that values new technology from the moment a researcher's instrument is installed—even going as far as encouraging researchers to bring their own instruments as an investment in strengthening the overall technical capability of the site.

5. Greater impact may be achieved by broadening data accessibility.

Observatories can increase their impact by extending the availability of the data collected at the site to a larger potential user base—thus increasing the overall value proposition of the observatory. Observatories that have prioritized data storage and sharing have seen an overall increase in publications generated from the data collected on-site. More publications can result in greater prestige for an observatory, ultimately expanding its potential customer base. Effective data archiving and sharing also allows researchers who have not been granted telescope time an opportunity to make valuable scientific discoveries using data collected by others.

AFRL should look into the resources necessary to modify its archiving system to preserve *all* data collected on-site for future access, not solely data that have been specifically requested. As experience at other observatories has shown, a policy of data sharing can greatly extend the impact of an observatory—even in facilities that deal in a significant amount of sensitive data. Data sharing can increase the customer base by attracting sensitive government customers who might be potential users of AMOS data. That said, AFRL should weigh the costs of such a system against the potential benefits to customers and the impact gained through robust data preservation and sharing.

The best practices presented here constitutes part of a larger research effort to determine AMOS's value proposition and mission moving forward, and how it may deliver necessary services at best prices while maintaining a budget that will reliably sustain operations. The benefit of these best practices could also extend beyond AMOS to other civil observatories struggling to streamline operations and maintain sustainable budgets.

Acknowledgments

We are very grateful to our U.S. Air Force sponsors, Dr. David Hardy and Col. Scott Maethner (AFRL Directed Energy Directorate), who were supportive of this work from the start.

This research would not have been possible without the contributions made by a number of individuals throughout the civil astronomy community. Mike Maberry at the University of Hawaii provided early guidance that helped shape our research plan, and we are thankful for all of the contacts that he helped facilitate to enable our data collection. Erin Elliott at the Space Telescope Science Institute (STScI) facilitated key introductions related to the Hubble Space Telescope (HST) and suggested resources that significantly impacted the analysis presented here. Dennis Crabtree offered valuable information contextualizing the output of observatories in the form of scientific publications. Finally, we thank all the astronomers who spoke with us about their own facilities; this research would not have been possible without their cooperation.

At RAND, Michelle D. Ziegler gave constructive insight and helped us refine our selection of case studies. Amy McGranahan provided steady administrative guidance over the course of the research period and was indispensable during the preparation of the initial manuscript.

The observations and conclusions made within this document are solely those of the authors, as are any errors or omissions, and do not represent the official views or policies of the U.S. Air Force or of the RAND Corporation.

Abbreviations

AFMC	Air Force Materiel Command
AFRL	Air Force Research Laboratory
AFSPC	Air Force Space Command
AMOS	Air Force Maui Optical and Supercomputing Site
ALMA	Atacama Large Millimeter Array
AO	adaptive optics
ATST	Advanced Technology Solar Telescope
DoD	Department of Defense
FFRDC	federally funded research and development center
FY	fiscal year
GEODSS	Ground-based electro-optical deep space surveillance (system)
HST	Hubble Space Telescope
ICD	interface control document
IRTF	Infrared Telescope Facility
LCOGT	Las Cumbres Observatory Global Telescope Network
MHPCC	Maui High Performance Computing Center
MIT/LL	Massachusetts Institute of Technology Lincoln Laboratory
NASA	National Aeronautics and Space Administration
NEO	near-Earth object
NOAA	National Oceanic and Atmospheric Administration
NOAO	National Optical Astronomy Observatory
NRAO	National Radio Astronomy Observatory
NSF	National Science Foundation
Pan-STARRS	Panoramic Survey Telescope and Rapid Response System
PI	principal investigator

PS1	Pan-STARRS (Panoramic Survey Telescope & Rapid Response System) 1
PS2	Pan-STARRS (Panoramic Survey Telescope & Rapid Response System) 2
R&D	research and development
SA	support astronomer
SSA	space situational awareness
SST	Space Surveillance Telescope
STScI	Space Telescope Science Institute
SWPC	Space Weather Prediction Center
TAC	time allocation committee
TMT	Thirty Meter Telescope
TO	telescope operator
UKIRT	United Kingdom Infrared Telescope
UH	University of Hawaii
WFPC2	Wide Field Planetary Camera 2

1. Introduction

Best Practices for Sustainable Operations at the Air Force Maui Optical and Supercomputing Site

This document presents RAND research dedicated to providing a slate of suggested best practices to the Air Force Research Laboratory (AFRL) intended to increase long-term sustainability and efficiency of operations at the Air Force Maui Optical and Supercomputing Site (AMOS). These recommendations and insights are intended to support the larger body of RAND research in service of AMOS, particularly an in-progress initiative to craft a value proposition and business plan for AMOS as its funding landscape changes.

The compiled best practices described here provide valuable lessons drawn from a broad spectrum of scientific enterprises. Although not all institutions we studied are directly commensurate—a civil observatory and a missile test range, for instance, do not face the same budget, staffing, and operational requirements as an Air Force-sponsored optical observatory—by collecting data from institutions with important attributes in common with AMOS, we have been able to assess a rich cross-section of experience and examples that will serve not only AFRL but also institutions facing common obstacles and common needs.

In order to frame the overall research project, the report begins with a brief overview of the AMOS facility itself, the services it provides to the Air Force's Space Surveillance Network, and some of the expected changes that have motivated the larger RAND study of AMOS.

AFRL operates a complex of three telescopes at AMOS, which is located at the summit of Haleakala on Maui. AMOS is unique in that it is the only Department of Defense (DoD) site capable of providing visible and infrared spectrum images of space objects passing over the Pacific Ocean. Together with three co-located telescopes run by the Air Force Space Command (AFSPC) as part of the Ground-Based Electro-Optical Deep Space Surveillance (GEODSS) network, the AMOS facility serves as a major site of U.S. space surveillance activity.

Each of the three AMOS telescopes provides unique resources to the complex as a whole. A 1.6-meter telescope conducts nonresolved photometry and can be used during the daytime. The 3.6-meter telescope is used for high-resolution imaging, and is the largest optical aperture within the DoD. The 1.2-meter telescope is currently not in use, but AFRL plans to bring it back into service for wide-field searches following an upgrade that will include rebuilding the mount, installing new hydraulics, and fitting it with a new sensor package.

When the 1.2-meter is brought back into service, its wide-field sensors will make it unique among the telescopes in use at AMOS and will broaden the overall technical capabilities of the site. Wide-field and narrow-field sensors have different limits and capabilities. Wide-field observations, such as those planned for the 1.2-meter, entail surveying large areas of the sky, imaging many objects of interest within a single exposure. Narrow-field observations, such as those currently underway at the 1.6- and 3.6-meter telescopes, zero in on specific objects and gather large amounts of data from those specific objects. Wide-field and narrow-field facilities rarely occupy the same site, so this dual capacity is part of AMOS's unique value proposition.

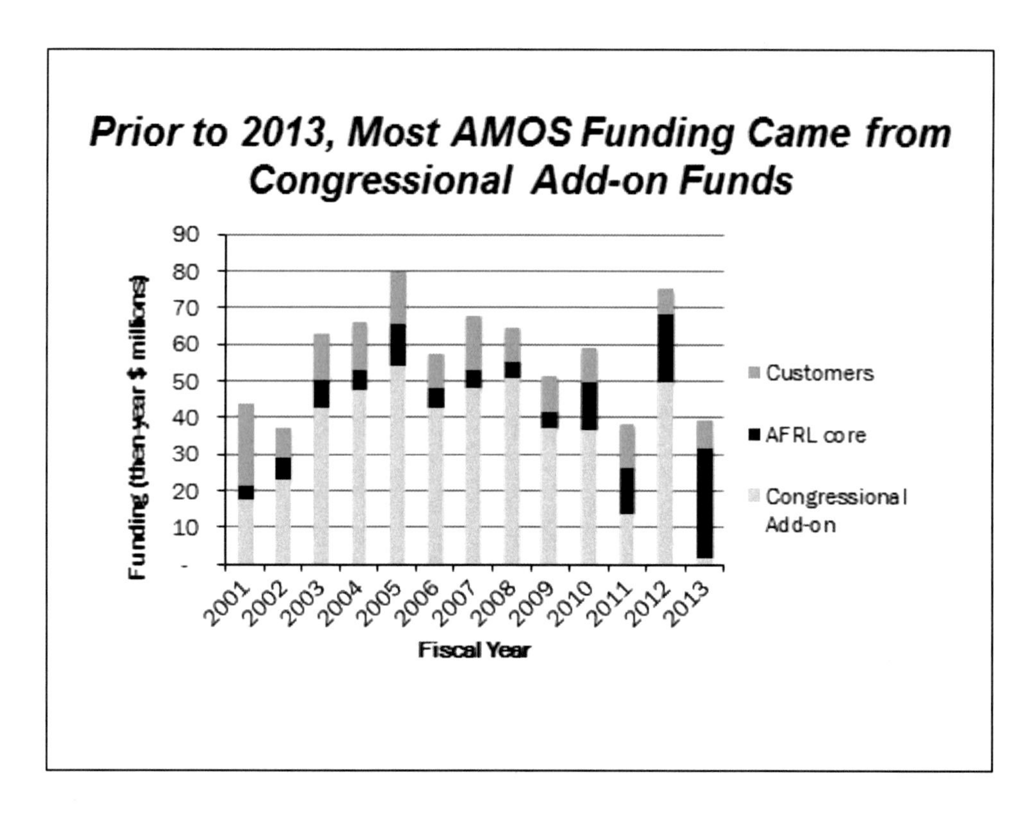

Since AFRL took over sole management of the site in 2001, the majority of funding used to run AMOS has been drawn from congressional add-on funds, which are also known as earmarks. This high level of congressional backing was largely the result of the support of Senator Daniel Inouye as an extension of his mission to establish science and engineering as a central part of the Hawaiian economy. While earmarks are no longer favored in the current budget environment, the support of Senator Inouye helped AMOS maintain a high level of congressional funding through 2012. As of fiscal year (FY) 2013, the level of congressional add-on funding has been greatly decreased and no longer constitutes the majority of AMOS funds.

As of 2012, congressional support remained the major source of money for the site. FY 2012 was the final year of congressional funding secured through Senator Inouye's support, and during this year AFRL increased its budgetary allotment to AMOS to support continued operations. A majority of the FY 2012 congressional add-on funds were dedicated toward upgrades to be completed over the next few years.

However, beginning in FY 2014 the site has been run under a new financial arrangement. To compensate for lost congressional funding, AFRL sought to lower its total yearly operating costs and drew up a memorandum of understanding (MOU) between itself and Air Force Space Command (AFSPC). Under this agreement, each party agrees to provide a set amount of funding each year to support the upkeep of facilities, equipment, and collection and delivery of data at AMOS, all of which continue to be the responsibility of AFRL. AFRL also holds responsibility for allocating observing time to all stakeholders and customers.

This Analysis Fits Into a Larger Body of RAND Research on AMOS

- **Phase I (completed June 2012): RAND assessed AMOS's (then) proposed modernization plan and evaluated the outcomes from previous studies of AMOS**
- **Phase II: Develop a value proposition and resulting business plan by looking at AMOS's historical context, best practices, customer base, and product lines.**

In anticipation of these changes in funding, Dr. David Hardy, Director of the AFRL Directed Energy Directorate, sponsored RAND to conduct two phases of analysis to benefit AMOS.

In Phase I, which was completed in June 2012, RAND assessed AMOS's (then) proposed modernization plan and evaluated the outcomes from previous studies of AMOS. This initial research assessed the Air Force's modernization plan to upgrade and repurpose the sensors in place at AMOS to meet key demands of space situational awareness (SSA), which refers to the ability to achieve and maintain comprehensive understanding of the position of natural and human-made objects in orbit. RAND evaluated AMOS's modernization plan and conducted interviews with the customer base, and found that the Air Force's plan aligned with customers' stated needs.

For Phase II, RAND is building on its initial review of existing AMOS customers to develop a value proposition and business plan intended to help the site achieve self-sustainability. This analysis is grounded in a detailed examination of AMOS's historical context, customer base, product lines, and potential best practices.

This document addresses the best practices component of the Phase II research plan. Over the course of several months, RAND worked to determine specific ways in which AFRL might operate AMOS more efficiently once the business plan has been implemented, and AMOS is operating steadily and stably under its new financial agreements.

Project Outline

- **How did we carry out this analysis?**
- **What are our key findings?**
- **What does this mean for AMOS?**

The remainder of this report is divided into three chapters. In Chapter 2, we describe our objectives and the methods used to carry out this analysis. Chapter 3 discusses the key findings we gathered from the accumulated insights and lessons learned over the course of the research period. We conclude in Chapter 4 with suggestions for how these key findings might be applied by AFRL in the process of making AMOS a more sustainable facility. We also provide a brief overview of how these findings might be put to use, both by AFRL and by other institutions facing changes in support.

Appendix A goes into greater detail on each of the case studies conducted over the course of the research period. Each case study description contains a wealth of information supporting the key insights presented in the main report. Additionally, these detailed case studies present insights and practices that extend beyond the scope of our main recommendations for AMOS, offering valuable considerations for similar research sites with concerns about staffing, scheduling, and funding. Appendix B provides useful perspective about how land leases for observatories located in Hawaii are negotiated—in particular how telescope time has been allocated differently at each Hawaiian observatory as part of land use agreements between the observatories and the University of Hawaii, which serves as land steward for many of the sites surveyed. It also contains a general overview of our points of contact at each institution surveyed. Finally, Appendix C contains a list of organizations that contributed data to this research.

2. Objectives and Methods

Our Objective Was to Collect Best Practices from Analogous Research Institutions

- **What factors support the efficient, productive, and sustainable operation of a government-run observatory with multiple scientific facilities on Hawaii?**
- **What can AFRL learn from civil observatories, military test sites, and research facilities employing a range of funding structures, operational approaches, missions, and value streams?**

Our objective for this research was to provide AFRL with a suite of suggestions for methods that might help it move toward greater self-sustainability following the implementation of the business plan developed at RAND. The key goal was to identify ways AMOS might streamline its R&D and operations missions while extending its impact and value to potential customers.

We sought to collect and distill a set of best practices from research institutions that have attributes in common with AMOS. We initially set out to identify institutions with at least one of the following three analogous characteristics:

- Observatories and other sites of research and data collection located in Hawaii.
- Government facilities in remote places that might provide insight on staffing and logistical matters.
- Research facilities that conduct blended operations and R&D.

We initially set out to locate institutions that met one or more of these attributes among civil observatories, government testing facilities, and research institutions across the U.S. government scientific enterprise.

We Began By Compiling an Initial Set of Key Attributes to Serve as Selection Criteria

- Blended R&D and operations
- Heterogeneous mix of expertise in workforce
- Need for agile and responsive management
- Part of a larger observing network
- Funding from multiple sources
- External customers as part of funding structure

In addition to the three primary criteria listed on the previous page, we identified a set of second-order qualities that grew out of these main considerations. We looked for institutions whose workforce included a high level of diversity—of background, education, and expertise. We sought to identify sites with a high level of technical complexity requiring an agile and responsive management structure, and sites that exist as part of a larger research network. We also identified institutions whose funding draws from multiple sources, including customers external to the primary sponsor or sponsors.

We did not seek out sites that had all these characteristics in one place, because such a place does not exist. Instead, we sought to sample a cross-section of institutions that could provide insight into all of the key attributes in aggregate.

We Compiled a List of Institutions That Possess a Cross-Section of Relevant Attribute

	Keck	Gemini	Subaru	NASA IRTF	LCOGT	VLBA/NRAO	Pacific Missile Test Range	Tsunami Warning Center	MIT/LL radars	SWPC NOAA	National Solar Observatory	SST	Kwajalein Test Range	NOAO HQ
Mission blend: R&D and operations	✓			✓			✓		✓	✓			✓	✓
High workforce heterogeneity	✓	✓	✓				✓		✓				✓	
Management structure of interest	✓	✓	✓						✓		✓			✓
Part of a larger network			✓		✓	✓		✓	✓			✓	✓	✓
Funding models of interest	✓	✓		✓	✓		✓		✓	✓	✓		✓	✓
Multiple customers	✓	✓	✓		✓		✓						✓	✓

To ensure a broad selection, we developed a framework that helped us generate an initial set of organizations that could represent the desired cumulative range of attributes. We cast a wide net and populated the framework with candidate sites that met one or more of the key criteria and would ensure broad representation of each attribute. A significant number of the initial set included sites in Hawaii—the leftmost eight institutions in the chart presented here have holdings on one of the Hawaiian Islands. We also included military test sites and a variety of national research organizations.

The chart in this slide represents our initial slate of candidate sites, which served as an example of the range of distribution we wished to attain. By making sure that each criterion was well represented, we could ensure that the institutions surveyed would be sources of information potentially useful to AFRL. We did not contact all institutions represented here; however, over the course of the research period, we adjusted the candidate sites as necessary to ensure that the distribution remained robust.

We Used a Standard Set of Questions to Guide Our Discussions

- Interviewee's professional biography, title
- Institutional value proposition
- Institutional makeup: management, mission
- Funding structure
- Type of scheduling, method of time allocation
- Division of time for maintenance, engineering
- Data management: storage, accessibility
- Level of remoting capability
- Support services: water, electricity, information

Before embarking on site evaluations, we established a protocol for data collection. Our research plan entailed detailed interviews with representatives from each site conducted by two RAND researchers who took notes over the course of each interview. We compiled detailed transcripts following each interview and evaluated the transcripts individually to determine the key takeaways from each site. Following completion of the interview phase of research, we assessed the cumulative transcripts and identified common themes and insights applicable to operations at AMOS.

In order to impose standards on the data we collected, we used a set of initial questions to guide each discussion, with the understanding that these questions might spin off in other generative directions. Some of the questions were intended to gather basic information—for example, the interviewee's professional biography and current role at his or her home institution; what he or she saw as the site's value proposition; and an overview of the institutional mission, management, and funding structures. We started with these questions in particular based on the primary concerns at AMOS identified in the results of Phase I research.

After gathering this general information, we went into greater detail regarding the technical specifics of each site. We examined how time on available equipment is scheduled and how maintenance and engineering fit into the schedule, given that downtime for technical necessities often means less time that could otherwise be used for active data collection. We inquired about data collection—if and how data collected on-site are processed and stored; if the data are made more widely accessible when stored; and, if so, who has access to archived data. We investigated the procedures in place at observatories that conduct remote observations. These sites have the ability to collect data from a location physically removed from the primary sensor, be that location at the bottom of the mountain whose summit houses the telescope or an ocean away on the mainland. We also sought to determine how institutions on Hawaii and other isolated locations meet infrastructure needs. While Hawaii might not seem like a remote location, these observatories are located high up on the summit of large, inactive volcanoes, where even water must be trucked up. Support services become a major consideration when certain types of infrastructure are lacking; since AMOS is located in one such remote location, these considerations were incorporated into the overall interview process. In discussing remoting procedures as well as infrastructure and information sharing, we sought to determine how facilities in remote and therefore expensive locations reduce operations costs.

We Conducted Interviews with an Expanded Set of Relevant Institutions

- W. M. Keck Observatory
- Panoramic Survey Telescope and Rapid Response System (Pan-STARRS)
- Kwajalein Missile Range
- Thirty Meter Telescope (TMT)
- Hubble Space Telescope (HST), Space Telescope Science Institute (STScI)
- National Optical Astronomy Observatory (NOAO)
- Gemini North
- MIT Lincoln Laboratory
- Los Cumbres Observatory Global Telescope Network (LCOGT)
- United Kingdom Infrared Telescope (UKIRT)
- Advanced Technology Solar Telescope (ATST)
- Subaru Telescope
- Space Surveillance Telescope (SST)
- NASA Infrared Telescope Facility (IRTF)
- Atacama Large Millimeter Array (ALMA)

After developing our research protocol, we determined a final list of sites to be surveyed. We identified contacts at these institutions based on prior meetings at conferences, introductions by third parties, and connections made by local administrative staff. As we conducted interviews, we made further connections with representatives of a larger set of relevant sites. Our initial set of interview targets grew into the list presented here. With each of these case studies, we spoke with one or more individuals who had intimate knowledge of how the sites are managed, day-to-day procedures, and a larger view of what unique services the site in question provides to its intended customers—whether civil astronomers, government agencies, or private entities. We typically interviewed individuals in leadership positions, many of whom also work as researchers or support staff at their respective institutions. None of the sites listed provided an exact, point-to-point analog to AMOS, but each gave us valuable insight regarding how to run an observatory under the unique conditions in place at AMOS.

3. Key Findings

Project Outline

- **How did we carry out this analysis?**
- **What are our key findings?**
- **What does this mean for AMOS?**

After completing all interviews and compiling the results, several key issues emerged as common concerns among the site representatives we interviewed. We compiled these recurring issues and conducted a synthetic analysis with the intent of building a suite of suggestions and insights that could best be put into practice at AMOS following its modernization effort and funding reorganization.

This chapter outlines the key findings on best practices compiled from our analysis of the cumulative case studies. The next chapter describes what these findings mean for AMOS and how they might inform future practices to move toward greater sustainability.

We Identified Five Key Factors That Will Affect How AMOS Can Provide Value in the Future

- Narrow-field vs. wide-field observing: differences in time management, staffing, data value
- Fully remote vs. on-site operations: differences in staffing, infrastructure requirements
- Messaging: Better outreach means wider understanding of institutional value, greater support
- Configuration flexibility: Agility of instrument swaps
- Access to data: Wider availability enables higher production

Of all the issues we discussed with the representative institutions, the five key factors listed in this slide stood out as the most relevant to AMOS's future sustainability and success. We summarize each point below and go into further detail in the upcoming slides.

First, we determined that important differences exist between telescopes that gather data from a wide field of view and those that focus on a single target at a time. These differences show up in staffing, time management for operations, engineering and maintenance, and the ways in which data from either type of observation can be used.

Second, we identified several differences between sites that are fully equipped to enable remote observing and those that require on-site staffing. "Remoting" refers to telescope and instrument control conducted from a site physically removed from the equipment itself, whether from the base of a mountain whose summit houses the telescope or from further afield. Based on these observed differences, we compiled a recipe for successful remoting that we describe in detail in the coming chapter.

Third, we observed that thoroughness of messaging, as an extension of an institutional mission that values internal and external outreach, can have a significant impact on breadth of support, both financial and cultural.

Fourth, we observed that maintaining flexible configuration can raise the value proposition of an observatory but requires either robust scheduling or the ability to quickly swap instruments.

And fifth, we observed that wider access to data yields higher knowledge production levels per amount of data collected, thus increasing the impact of the institution collecting the data.

We Observed that Wide-Field and Narrow-Field Observatories Are Run Differently

- Multiple users can exploit the same rich data set, increasing knowledge production by a single telescope
- Survey data enables *in situ* engineering
- Professional surveyors on staff rather than generalist astronomers

Our first key observation identifies several important differences in operations at observatories that conduct wide-field observations and those that observe in the narrow field. Because AFRL intends to manage both types of telescopes at one site, these differences are of particular interest with regard to the different ways that the 1.2-meter should be run as compared to the 1.6- and 3.6-meter.

By their nature, **wide-field observations are extremely rich in data**. Multiple targets are collected at once, rather than one at a time as in targeted narrow-field observations. Although narrow-field observing may collect more detailed data from a single object during a single observation, a significant amount of data about multiple objects can be gleaned from a wide-field exposure. This often results in a larger number of users who might be able to use data collected during a single wide-field observation period.

Another advantage of wide-field observing is that, unlike narrow-field observing, which requires telescope downtime for engineering**, data taken during wide-field surveys provide *in situ* feedback on how well the instruments operate**. Comparing data on the same object taken on different nights under different atmospheric and technical conditions enables comparison and adjustment without setting aside telescope time specifically for engineering tasks—time that could otherwise be used by researchers.

We also noted that **those telescopes used for wide-field observations benefited from having professional surveyors operate the telescopes**, rather than the generalist astronomers who typically supervise operations at narrow-field telescopes. The facilities run by Pan-STARRS in Hawaii provide an excellent example of this staffing model. Part of this specialization is enabled by the repetitive nature of survey astronomy missions—in performing the same type of observation on a regular basis, these staff astronomers become specialized technical experts in the art and science of sky surveys. Employing specialized wide-field observers for survey operations can lead to less downtime and more-streamlined scheduling.[1]

[1] For more on the unique scheduling and staffing possibilities of wide-field observing, see Pan-STARRS, Appendix A.

Some of the Facilities Surveyed Lowered Overhead Via Remoting

- **Strong remoting capabilities may reduce the cost of infrastructure needed to support visiting researchers (lodging, transportation, etc.), but we also found evidence that these cost savings may be transferred elsewhere**

- **Two different flavors of remoting**
  - **Summit to base**
  - **Summit to base to Continental US**

Irrespective of the type of data collection under way at each site, we noted the expanding practice of remote observation at several key observatories. In the absence of cost data, we saw evidence to suggest that a robust remoting infrastructure allowed some observatories to lower their overhead costs by cutting out services in place to support on-site researchers, such as lodging, transportation to and from the summit, food, and other basic needs.[2] However, in some cases, especially at those observatories that require hands-on access to the instruments, the costs saved in cutting these services were transferred directly to increasing support staff on-site, particularly support astronomers who assist researchers observing from remote locations. Instead of paying travel costs to bring a principal investigator (PI) to visit the facilities in person, a salaried staff scientist serves as a scientific consultant and technical liaison for the absent PI.

That said, some observatories that have established remoting programs have managed to reduce the number of staff on summit at night to either two or zero—we learned that if there is one person in the dome there must be two, in case of emergency. Those observatories with no staff directly on-site must have instrumentation that can swap itself without human intervention, or can be set during the day. For observatories that wish to maximize available observing time, this represents a significant upfront investment—either in technical upgrades to the telescope mount or in developing and implementing improved scheduling procedures that enable flexible time allocation.[3]

[2] Per phone conversation with personnel at Herzberg Institute of Astrophysics, July 1, 2013.

[3] See NASA IRTF and Gemini, Appendix A.

Two types of remoting characterize those observatories that do not send staff to the summit at night. Some remoted observatories conduct observations and control the telescope from control rooms at the base of the mountain whose summit houses the telescope and instruments. Observatories that allow PIs from further afield to observe from sites closer to home have set up technical specifications and standards that enable the telescope to be controlled from a remote center at the base of the mountain, and the instruments to be controlled by researchers from the continental United States (in the case of those observatories located in Hawaii). These two different levels of remoting require different technical setups and staff, depending on the level of security desired by the observatory and the extent of control afforded the remote observer.

Fully Remote Operations Require Installation of Several Foolproof Systems

- **In addition to instruments that can be day-mounted and automatically switched, a fully remoted observatory must have:**
  - **Steady stream of telemetry**
  - **Ability to open and close telescope**
  - **Reliable, accurate method for measuring weather conditions at the dome**
  - **Standard tech specs for PIs**
- **Adaptive optics (AO) systems are difficult to remote**

In addition to reliable instrument switching capabilities, we determined that remoting can provide cost savings when a reliable set of standards and protocols is strictly enforced. A number of systems and practices must be in place to support both types of remoting in the absence of a full on-site staff.

- A steady stream of telemetry must be supplied by detectors and actuators situated at all points of possible failure, including at pieces of equipment that are largely dormant.
- Remote staff must have ready access to this telemetry so as to be aware of any problems that arise, particularly those that require staff to visit the site for repair.
- The telescope dome must be able to be closed or opened remotely based on weather changes.
- Weather changes must be readily measurable from afar.[4]

In order for a remote visiting researcher to be actively involved in data collection, he or she must adhere to a set of strict technical specifications set by the observatory in order to reduce the incidence of error. Keck Observatory and the NASA Infrared Telescope Facility (IRTF) are both examples of observatories deeply invested in remote observing. Keck Observatory provided us with a particularly strong set of remoting interface control procedures that staff use to ensure that all remote PIs adhere to a strict protocol designed to ensure efficient observing time and secure data transmission. These documents outline the technologies a PI must have in place at a remote observing site, such as operating systems, software, and display resolution, required power

[4] For suggestions on weather data sharing on Mauna Kea see Pan-STARRS, Appendix A.

supply specs and tech support requirements, and even the layout of the remote observing room. If any of these specs are not met—or if they fail—Keck staff is not responsible for conducting observations in place of the remote observer.[5]

IRTF allows a more flexible degree of remoting than Keck in that any approved observer with an Internet connection may control the instruments from afar. This occasionally results in lost data due to drops in Internet connectivity and raises the possibility of data interception, but not to the extent that IRTF has had to alter its relatively less rigorous model of remoting.[6]

A major caveat to these requirements currently exists for those telescopes that run adaptive optics (AO) systems. Atmospheric turbulence causes stars to twinkle, and AO systems subtract that turbulence and remove the twinkle for sharper images. AO systems are currently too temperamental to be operated remotely and require on-site caretaking. This is relevant to AMOS because the 3.6-meter telescope has AO. Remoting may thus be more difficult or impossible for the 3.6-meter at this time, so any potential remoting initiatives should focus on the 1.2- and 1.6-meter telescopes.

---

[5] Keck maintains a set of publicly available specifications for observing from the mainland. See http://www2.keck.hawaii.edu/inst/mainland_observing/ (accessed 9/17/2013).

[6] For more detail on remote observing at IRTF, see NASA IRTF, Appendix A.

Greater Impact May Be Achieved by Broadening Data Accessibility

- **Wider data availability enables more publications**
- **Gemini, UKIRT, Keck, Pan-STARRS, ALMA, and HST prioritize data access**
- **UKIRT saw a 4x yearly increase in publications produced using UKIRT data after sharing policies were changed to make the data more widely accessible**

We received anecdotal evidence that dedicating staff and resources to outreach can result in greater support both internally and externally—and that both internal and external messaging is key for maintaining a successful R&D facility.

MIT Lincoln Laboratory (MIT/LL), in particular, illustrates the value of good internal messaging. In speaking with representatives of their radar facilities and the Space Surveillance Telescope (SST), we learned that comprehensive internal messaging has been an integral part of maintaining steady funding. All staff members, from technicians to managers, can discuss the technical advantages of the site accurately and are capable of tying these advantages back to benefits for the sponsor. As a result, the value proposition of the entire site is well known and held up as part of the fabric of the institution.[7]

The Space Telescope Science Institute (STScI), which manages the science mission for the Hubble Space Telescope (HST), is an exemplar of outreach and external messaging, with a staff division dedicated to publicizing Hubble's accomplishments and to keeping its contribution to astronomy in the public mind and, by extension, supported by legislators. Their effectiveness at outreach is easily observed: Most of the general American public is aware of HST, and many of the discoveries it has enabled have reached iconic status within American culture.[8] While the outreach model used by STScI cannot be directly applied to facilities with sensitive customers, we suggest that a strong program of external outreach, even within constrained parameters—

[7] See MIT/LL, Appendix A.

[8] See, for example, the "Pillars of Creation" image of the Eagle Nebula, Hubble Deep Field, and Hubble Ultra Deep Field, among others. HST itself is also the subject of a widely distributed IMAX film.

keeping those who can affect funding consistently in the know about a site's value and contributions—can have a positive impact on the consistency of support among stakeholders in a variety of scientific institutions.[9]

[9] See STScI/HST, Appendix A.

Flexible Configuration Can Lead to Downtime, but Also Added Value

- **Allowing visiting PIs to bring own instruments means extra downtime to configure new instruments and recalibrate equipment to base configuration**
- **But: equipment left behind by PIs can provide added value through**
  - **Expanded technical horizons for future researchers**
  - **Larger pool of potential customers**

At many observatories and research sites, including AMOS, fielding instruments brought by visiting PIs has often been seen as a nuisance. These instruments may require downtime to set up and money to maintain after the PI departs.[10]

However, those institutions we surveyed that had a strong R&D mission, particularly the Space Control Research Group at MIT/LL (which oversees the operation of the SST, for example), emphasized the value in allowing PIs to bring instruments. MIT/LL sees this as expanding the ways in which the facilities may be used—more instruments means more types of research that can be conducted on-site. Combined with effective outreach, as mentioned earlier, SST has been able to attract researchers interested in using data collected with instruments brought to the facility.[11] In addition, civil observatories such as Subaru have built the capability to field visiting instruments into the telescope itself by designating two of ten available instrument slots as dedicated "PI instruments."[12]

Whether the instruments are being developed at universities for use at civil observatories or at government research facilities for use at a site like AMOS, allowing flexible configuration can add significantly to an observatory's value proposition by expanding its potential customer base.

---

[10] For more on the scheduling and budgeting conflicts that often characterize flexible configuration, see TMT and MIT/LL, Appendix A.

[11] See MIT/LL, Appendix A.

[12] See Subaru, Appendix A.

Greater Impact May Be Achieved by Broadening Data Accessibility

- **Wider data availability enables more publications**
- **Gemini, UKIRT, Keck, Pan-STARRS, ALMA, and HST prioritize data access**
- **UKIRT saw a 4x yearly increase in publications produced using UKIRT data after sharing policies were changed to make the data more widely accessible**

We identified several observatories that have broadened their impact by making data collected on-site more broadly available to a larger potential user base—thus increasing these observatories' overall value proposition. Several observatories that have prioritized data storage and sharing have seen an overall increase in publications generated from data collected on-site. The archived data is made available—either public or to a constrained set of potential users—usually following an embargo period in which only the initial PI has proprietary access to his or her data. This increases the impact and efficiency of an observatory while providing some measure of intellectual property protection for the PI. Once the embargo is lifted and a larger audience obtains access to the same data, a greater number of papers may be produced by a larger number of data users. Data archiving and sharing provides potential dividends beyond papers produced by those who have gone through lengthy proposal and approval process and allows those not granted telescope time the chance to make valuable scientific discoveries using data collected by others.

Gemini, UKIRT, Keck, Pan-STARRS, ALMA, and HST, in particular, prioritize data accessibility using different types of archive systems and different constraints on access.[13] For example, UKIRT saw a fourfold yearly increase in publications produced using UKIRT data after its sharing policies were changed to allow access by the entire UK community following the standard one-year embargo.[14]

---

[13] See Gemini, UKIRT, Keck, Pan-STARRS, ALMA, and STScI/HST, Appendix A.

[14] See UKIRT, Appendix A.

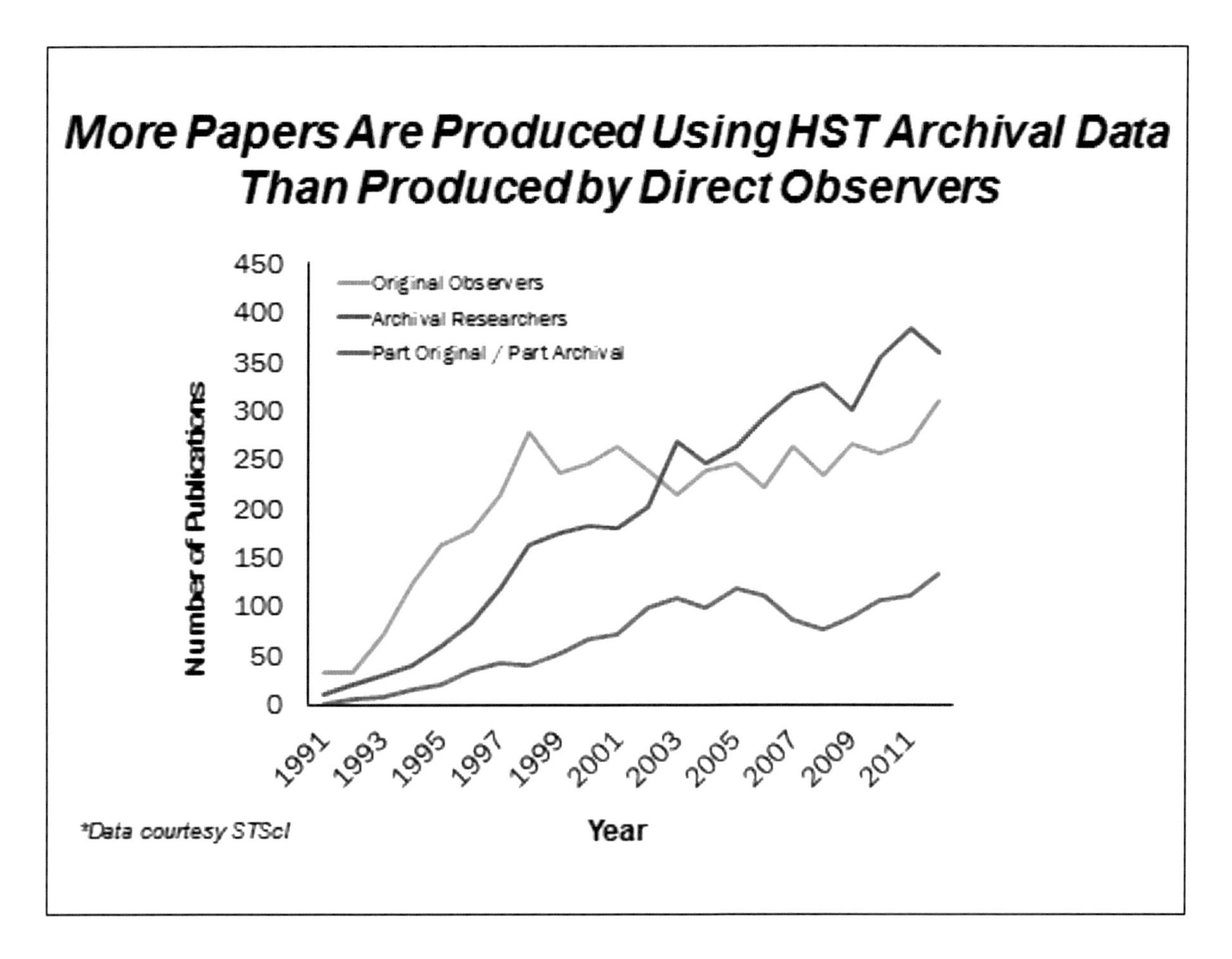

HST and STScI provide an exemplary model for how a well-catalogued, stable, open-access database can pay large dividends in overall impact of a data-collecting facility in terms of knowledge production.[15] STScI maintains a searchable database accessible to nearly everyone in the world, providing open access to HST data following a yearlong embargo during which only the PI who requested and was awarded telescope time can use the data he or she collected. This chart shows the number of publications produced using data collected by PIs granted observing time on HST. The green line represents publications by people who won time on the Hubble and had initial proprietary access to the data they collected. The purple line represents publications by people who did not have time on Hubble, but who waited out the yearlong embargo and made discoveries based on data in the archive. The blue line represents papers that used both initial and archival data, demonstrating that archival data are not merely useful to those researchers who are not awarded telescope time. Archival data can be valuable also to researchers who do get their proposals accepted and collect data directly.[16]

[15] Selected data from "HST Publication Statistics," Space Telescope Science Institute, accessed September 17, 2013, http://archive.stsci.edu/hst/bibliography/pubstat.html.

[16] For more detailed information on the types of publications produced by HST based on its data sharing model and available instruments, see STScI/HST, Appendix A.

This open access model has had a clear impact on the amount of knowledge produced from Hubble data. Since 2003, the number of papers published in refereed journals using HST archival data has surpassed the number of papers published by observers who requested the original data collection. This suggests that archived data can serve as a rich source of information and, in the case of HST, can significantly increase an institution's value proposition.[17]

In addition, the green line representing papers published by PIs using proprietary data has somewhat leveled off since 1998. This is likely due to the fact that there is only so much observing time available on Hubble within technical constraints. A fixed number of primary observers get to use Hubble data, whereas an unlimited number of researchers can make new discoveries using archived data.[18]

In addition to the five key findings, several second level lessons could provide added value to AMOS as AFRL considers how best to optimize operations in the future. While not all of these lessons will directly transfer given AMOS's unique funding and administrative structure, as well as the sensitivity of the site, these institutions provide excellent models of successful programs to manage such relevant concerns as automated scheduling, labor division, and hardware control. The next chapter demonstrates how the five key factors may be integrated at AMOS but does not go into granular detail about these second-level lessons. Table S.1 on page viii of this document connects the second-level lessons to the institution or institutions surveyed that provide potentially useful information on each lesson. For more details, see the institutional profiles compiled in Appendix A.

---

[17] Publications in scholarly journals are not the only way that widely accessible data may increase an observatory's impact. The Hubble Heritage Project provides an excellent example of archive users (not all of whom are trained astronomers) who have used Hubble's open data to create visually compelling images that have raised the public profile of HST.

[18] For more detailed information on the types of publications produced by HST based on its data sharing model and available instruments, see STScI/HST, Appendix A.

4. Implications of Key Findings for AMOS

Project Outline

- **How did we carry out this analysis?**
- **What are our key findings?**
- **What does this mean for AMOS?**

This final chapter addresses each of the key findings described in the previous chapter and explains their implications for AFRL. These suggestions will serve to increase efficiency of operations, particularly at the currently mothballed 1.2-meter telescope slated to be brought back online for wide-field observing, as well as increase the overall impact of AMOS data through streamlined administrative and technical protocols, stronger outreach, and an institutional mission that caters to a larger, more diverse customer base.

AMOS Should Operate the 1.2m with Unique Wide-Field Survey Attributes in Mind

- **AFRL needs to recognize that the 1.2m will need to be run differently than the 1.6m and 3.6m**
  - ***In situ* maintenance means less downtime**
  - **It is preferable not to switch staff between mounts**
- **Rich data should be saved for later use by other researchers**

As discussed earlier in this document, AFRL plans to bring the 1.2- meter telescope back into use from its current mothballed state to conduct wide-field area searches. This type of observing constitutes an entirely different mission from the two currently active telescopes on site. In bringing the 1.2-meter back online for wide-field observing, AMOS should be mindful of the differences between wide-field methods and the specific, narrow-field observations currently being conducted using the 1.6-meter and 3.6-meter telescopes.

In particular, AFRL should be aware that scheduling at the 1.2-meter will be different than the scheduling currently in practice at the 1.6- and 3.6-meter telescopes. A significantly smaller portion of total telescope time will need to be devoted to engineering and calibration due to the unique ability of wide-field survey observations to facilitate *in situ* observations. AMOS should take advantage of the unique attributes of the wide field of view—as Pan-STARRS has—to use data collected during observations to help inform calibration and engineering needs.[19] This is an advantage that wide field of view provides over narrow-field observations, but AFRL will need to plan ahead for this difference in order to realize the potential benefits.

AFRL should also consider employing "professional surveyors" to run the 1.2-meter, rather than switching staff regularly between the different telescope mounts on-site. While generalist astronomers can effectively operate both wide-field and narrow-field telescopes, we observed that having staff dedicated to wide-field missions produces experts in the observational practices

[19] See Pan-STARRS, Appendix A.

unique to wide-field observing. A degree of efficiency may be gained by making this designation in observing staff.

Finally, survey observations, by definition, are rich with data because they collect data over large swaths of the sky. The raw data collected by the 1.2-meter should be preserved for later use by potential customers, regardless of the initial data needs of the original PI. This practice may require some raw computing infrastructure, but it has the ability to increase AMOS's ability to poll historical data for change detection when addressing customer issues. We illustrate this point further by providing an example from HST on the next slide.

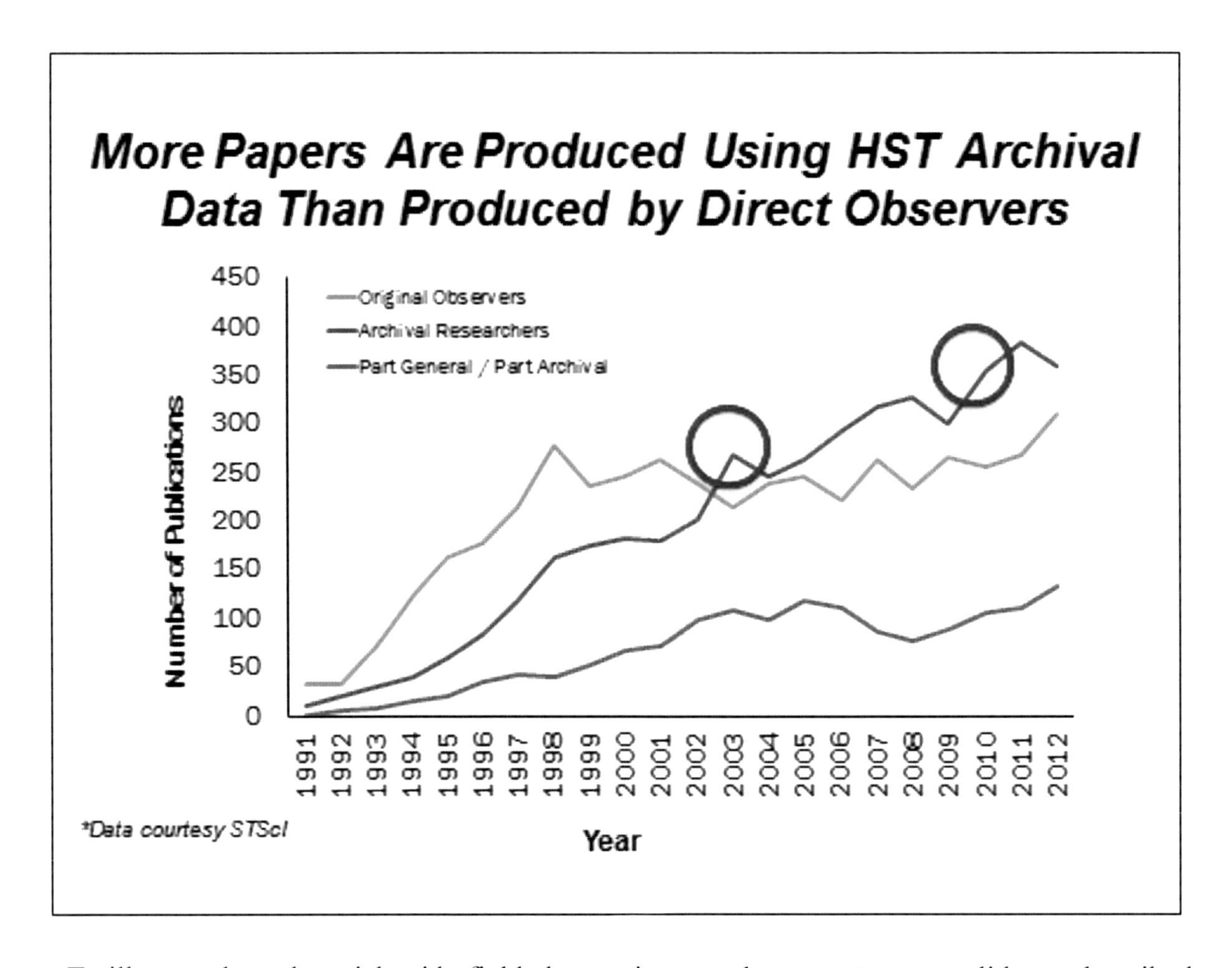

To illustrate how data-rich wide-field observations can be, we return to a slide we described earlier to highlight the uptick in archival publications (represented in purple) that took place in 2003. What caused this uptick? In 2002, a new wide-field survey camera was installed in Hubble. We hypothesize that the sharp increase in publications during this year corresponds to publications that would have been produced after the embargo on 2002 data was lifted, representing an increase in valuable data useful to astronomers worldwide. The number of publications produced by archival data continues to rise steadily following this initial spike. We see a similar uptick in 2010, following the May 2009 installation of another wide-field camera in Hubble.[20]

This plot suggests a connection between wide-field instrumentation and an increase in data output. On the next slide, we will take this a step further by suggesting that adding wide-field instruments to a telescope can also lead to an increase in the amount of data that a site can mine for additional information.

[20] Selected data from Space Telescope Science Institute, "HST Publication Statistics."

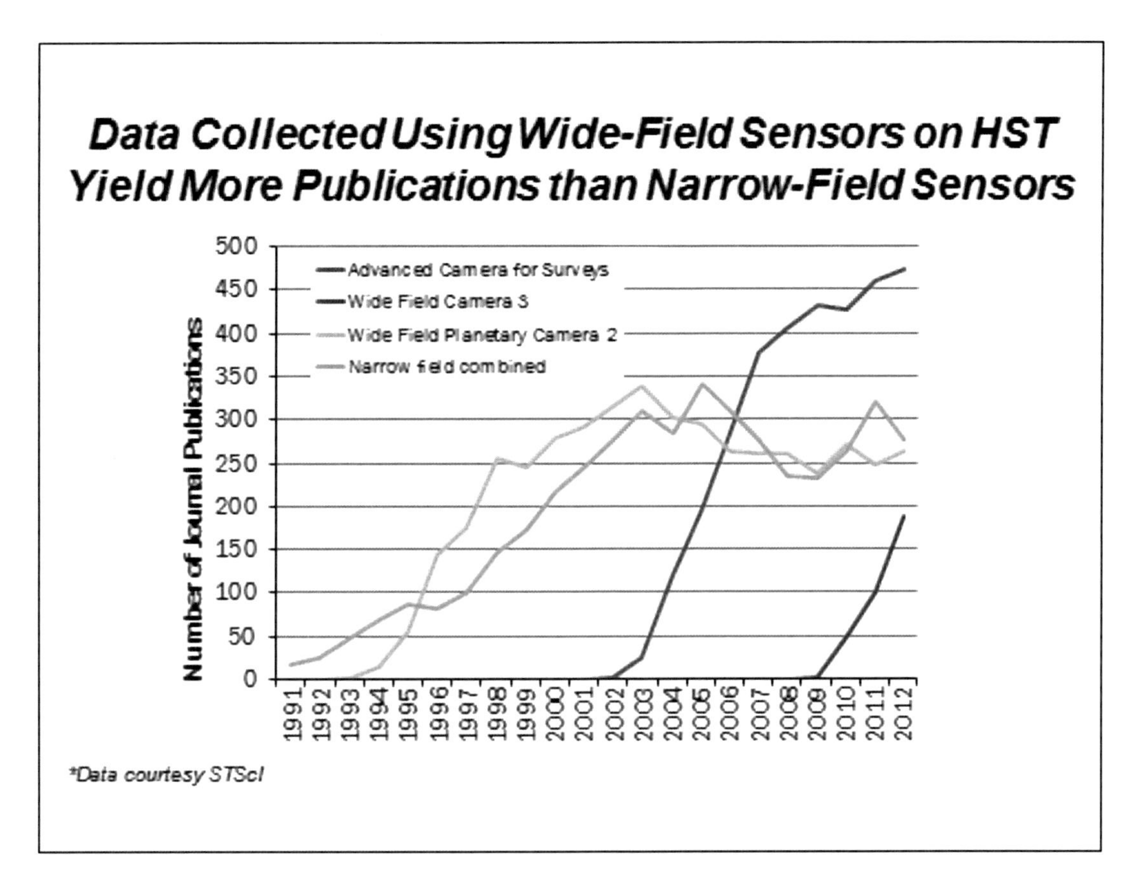

This chart illustrates the difference in publications from data gathered by wide-field Hubble instruments and data gathered by narrow-field instruments. Hubble's two currently operating general wide-field cameras (represented in red and blue), installed in 2002 and 2009 respectively, have seen sharp, continuous increases in publications whereas publications using data gleaned from Hubble's narrow-field instruments (represented in aggregate in green) have leveled out over the past decade. The difference in publication rate between papers written using data from wide-field and narrow-field instruments on HST illustrates how rich wide field observations are in terms of data, and how this data can increase the scientific impact of an observatory.

The Wide Field Planetary Camera 2 (WFPC2)—represented here by the orange line—provides another interesting case study in the utility and richness of wide-field data, as well as the importance of a robust data storage and sharing plan in increasing data impact. WFPC2 served as Hubble's main camera until superseded by the Advanced Camera for Surveys in 2002, and was removed and replaced with the Wide Field Camera 3 in 2009. In the years following its removal from Hubble, WFPC2 has had a notable afterlife in publications produced using archived data collected by the sensor during its operational lifetime. The number of publications that use archived WFPC2 data remains comparable to the number of publications produced from the total data yield of all narrow-field instruments combined.[21]

These data from HST illustrate the potential usefulness of wide-field data, but what can AMOS learn from this insight, especially because AMOS's primary objective is not the writing and publication of scientific papers? First, just as the HST's value proposition to the astronomical community increased after improved wide field instruments were installed, it is reasonable to assume that AMOS will likely realize a similar increase after the twin 1.2-meter (wide-field) telescopes are brought online. However, a significant portion of the value derived from HST's wide-field data was mined from the HST data archive, and this suggests a second lesson for AMOS: In order to maximize the data from wide-field instruments, AMOS will need to collect and store all the data in a searchable archive.

However, it is worth noting that AMOS deals with more sensitive information than does the Space Telescope Science Institute, and further research will be needed to identify which implementation of data sharing will provide the most benefit to AMOS and its customers. Unlike the HST, AMOS serves many different users, and any data sharing arrangement will need to consider the sensitivities of this broad user base. In our discussions with AMOS, we learned that such data sharing policies have been in place in the past—demonstrating that it is possible to make such arrangements—but these policies will need to be revisited and updated to ensure that they are pertinent to the current stakeholders.

Provided that AMOS is able to put such policies in place, even the most sensitive Maui customers stand to benefit from a historical data repository. The reason is that maintaining custody is one of the biggest challenges in SSA, and having a historical record available will allow analysts to "go back in time" to analyze past behavior whenever a new object is discovered for the first time.

[21] Selected data from Space Telescope Science Institute, "HST Publication Statistics."

We Propose a Two-Step Method to Assess Whether or Not to Remote AMOS

1. **Evaluate which ingredients will be necessary to remote**

2. **Consider that full remoting will not necessarily mean a drop in labor—it will likely require _new types of labor_**

Remoting has the capacity for reducing costs in the form of lower staffing and infrastructure needs. However, remoting is not yet foolproof, nor is cost reduction guaranteed. We suggest that AFRL evaluate whether or not it should remote the 1.2- and 1.6-meter telescopes based on the following two-step method, keeping in mind that the AO systems on the 3.6-meter make it harder or impossible to remote at this time.

First, AFRL should assess which technical changes would be necessary to remote the site. The results of our interviews indicate that investments either in improved scheduling software or instrument-swapping technology can increase efficiency in remote telescope operation. To aid with this process, we suggest asking the following questions: What technical and staffing changes would be necessary to remote the site? How much would it cost to implement these changes, and would the eventual payoff in savings be worth the initial cost?

Second, in spite of these possible savings, AFRL should also keep in mind that remoting does not always guarantee savings in labor costs. Sometimes, it requires investment in different kinds of labor, not less labor. As we observed at Keck, IRTF, Subaru, and Gemini, how much and what kinds of new labor will depend on several factors.[22] These include the extent of technical changes installed to support remoting and the extent of support necessary to field customers, whether they come to the control site or conduct their observations from further afield. AFRL should consider these potential costs when determining whether or not to remote the AMOS telescopes.

[22] For different models of staffing based on levels of remoting, see Gemini, Keck, and Subaru, Appendix A.

AMOS Should Consider Maintaining a Strong Messaging Initiative

- **A strong messaging initiative may foster a higher awareness of AMOS's value proposition among partners and customers**
- **AFRL should cultivate a messaging program aimed toward specific audiences**
  - **Message machine: Updates to general government audience**
  - **Targeted messaging: Updates to existing and future customers**

Given the importance of strong internal and external messaging among both civil observatories and federally funded R&D sites in sustaining financial support, AFRL should consider cultivating a dedicated messaging initiative so that AMOS's unique contribution to U.S. SSA is broadly and thoroughly known inside and out of the Air Force.[23]

This can be done in two ways, both of which ideally should be implemented to build a comprehensive outreach program, even within the context of a sensitive site. To cultivate a strong general messaging initiative, AMOS might consider sending out a one-pager, pamphlet, or detailed e-mail twice a year to general officers and members of the senior executive service. Such a document should detail recent upgrades, notable successes, changes in staff or gains in expertise, new papers and publications produced using AMOS data, and the ways in which AMOS has provided value for the SSA enterprise. For targeted messaging, more individualized, case-specific documents could be sent out to various personnel engaged in the American SSA effort.

As suggested by representatives of MIT/LL, a strong, comprehensive internal understanding of mission and value proposition among AMOS staff, driven by mentoring and a cultural focus on mission awareness, can also bolster the appeal of an institution to potential customers.[24] To follow a similar model, AFRL should consider developing a method of educating all members of

---

[23] See MIT/LL, Keck, and STScI/HST, Appendix A.

[24] See MIT/LL, Appendix A.

the organization about the value proposition of AMOS, making sure that all jobholders can communicate how their role in the institution contributes to this value proposition.

If AFRL wishes to seek out new customers to help bolster its funding input, these messaging initiatives may go a long way toward solidifying AMOS's value proposition in the eyes of potential site users.

AMOS Management Should Prioritize Potential Value When Adopting Flexible Configuration

- **The value of AMOS's R&D capability would be strengthened by dedicated procedures to facilitate—and encourage—customer-brought instruments**
- **AFRL should approach PI hardware with a mindset of added value from the moment of arrival, not the moment of abandonment**

An R&D facility can benefit greatly from building flexibility into equipment configuration. Rather than strictly controlling setup and constraining available instruments to those owned and operated by in-house staff, we have observed that facilitating instruments brought by external researchers can broaden the institution's potential customer base, expand the technical scope of SSA missions, and diversify AMOS's value proposition.

Flexible configuration may cost more than a fixed configuration, in terms of both time and money. Allowing visiting PIs to bring their own instruments can mean extra downtime for configuration and recalibration to base configuration after the instrument is uninstalled, or extra costs to operate and maintain if the instrument is left behind by the PI. However, allowing and even encouraging PIs to bring their own equipment may open up use of the facility to customers who might otherwise go elsewhere to test and use specialized instruments.

Realizing this value requires a change of mindset. Should AFRL choose to revise its protocol for visiting experimenters, it could consider adopting an approach that values new technology from the moment the PI's instrument is installed. As part of this mindset, AFRL would consider encouraging PIs to bring their own instruments as an investment in strengthening the overall technical capability of the site.[25]

[25] See MIT/LL and Subaru, Appendix A.

AMOS Could Increase the Impact of Observations by Making Data More Accessible

- **The future of observational astronomy appears to be moving toward research produced through access to archived data, rather than data collection itself**

- **AFRL should consider developing a system to archive all data collected at AMOS, not just data useful to the PI**

As we illustrated by demonstrating the increase in publications produced using archival data among several case studies, observational astronomy appears to be moving into a new phase in which archival research plays a significant role in astronomical knowledge production.

At present, AMOS archives the data collected at the site. However, the preserved data includes only those data specifically requested by a particular researcher about a particular object. Other potentially useful data collected in the process of fulfilling the specific request is not preserved. The currently discarded data, if archived and made accessible, could prove useful to a larger number of potential users in a context external to the initial request.

Rather than discarding any data collected at AMOS not specifically requested by the PI, as it does now, AFRL should look into the resources necessary to bolster its archiving system to preserve all data collected on-site for future access. As STScI, UKIRT, Keck, and Gemini, among others, have shown, a policy of data sharing can greatly extend the impact of an observatory.[26] Even in the case of a sensitive site like AMOS, such a policy of data archiving and sharing within the U.S. government could attract several sensitive government customers that Phase I of the larger RAND AMOS project identified as potential users of AMOS data.

That said, the costs of expanding the archival system should be weighed against the potential benefits of customers and impact gained through robust data preservation and sharing.

---

[26] See STScI/HST, UKIRT, Keck, Pan-STARRS, Gemini, ALMA, and IRTF, Appendix A, for examples of observatories that have prioritized data storage and sharing. See UKIRT and STScI/HST, Appendix A, for data on the impact of data sharing on publication numbers.

The Findings From this Research Will Make an Impact in Two Ways

- **Best practices will provide specific suggestions for ways that AFRL could increase impact and streamline operations at AMOS**
- **Possible wider distribution for civil observatories with similar concerns**

These recommendations form the bulk of the best practices suggested for smoother operations at AMOS in the future. How does this fit into the larger RAND project for AFRL, and how do we see it being applied?

In the big picture, the research presented here informs the main Phase II research objective of determining AMOS's value proposition and mission moving forward, and how it may deliver necessary services at best prices while maintaining a budget that can reliably sustain operations. Once this first-order research goal is achieved, the lessons gained from analysis of relevant analogous institutions will be folded into the larger piece on sustainability practices.

In addition to this primary goal, the research presented here will also likely have a broader impact that extends beyond its initial sponsor. Over the course of several interviews with civil observatories, we received multiple requests that the information yield from this research be made available to civil observatories that are also struggling to streamline their shops. In making the open-source material of this best practices project available for access to the civil astronomy community, we hope that a larger group in need of such analysis but lacking the funds to support it might also benefit from the Air Force's investment.

Appendix A. Overview of Interview Insights

This report has presented a summary of the accumulated case studies in support of our broader, synthetic analysis. Over the course of several months of interviews, however, we collected a significant amount of information and insight beyond that used to support the key findings and suggested applications described in the body of this briefing. In this appendix, we provide a detailed overview of each of the sites surveyed and describe how the cumulative best practices gathered from each site could be adopted AMOS's benefit.[27] These overviews may prove useful to AFRL should future plans for AMOS intersect with any of the information collected and presented here.

[27] Interviews with AMOS leadership were beyond the scope of the current study. However, future researchers might include best practices gathered from personnel at AMOS itself.

SWPC Has a Unique Model for Combining Researchers and Observers at One Location

- SWPC does not run an internal R&D division, but instead allows researchers from the University of Colorado to use SWPC facilities for R&D through CIRES
  - Dialog between two groups in shared space facilitates mutually beneficial information exchange.
- Air Force could employ a similar model, fielding visiting researchers from national labs, service academies, and other government research sites

AMOS faces a challenge in its need to combine both operations and R&D. A similar challenge has been faced by the Space Weather Prediction Center (SWPC), which is the official U.S. source of space weather alerts, watches, and warnings. In 2005, SWPC moved from the Office of Oceanic and Atmospheric Research (OAR) to the National Weather Service (NWS), which is an operational division of the National Oceanic and Atmospheric Administration (NOAA). As a result of the move, SWPC's core mission changed from a primary focus on R&D to a primary focus on operations.

Although SWPC no longer has a research mission, it has an arrangement with researchers from the University of Colorado that allows the Center's facilities to be used for R&D. The R&D mission is carried out by 20 researchers from the University of Colorado through the Cooperative Institute for Research in Environmental Sciences (CIRES), which is a renewable cooperative agreement. Researchers are funded through external grants and other contract work. The arrangement offers benefits to both researchers and SWPC operators. Research staff benefit from the prestige of sitting inside NOAA, and the affiliation gives them an advantage when applying for research grants. SWPC benefits from having research PhDs on-site to bolster on-site R&D.

The scheme allows operations-focused individuals and researchers to work together effectively. A key is that they are all housed in the same building, which facilitates dialogue between the two groups. Forecasters are able to ask PhD-level scientists for guidance when they see something unusual happening. Scientists also have an opportunity to participate in forecasting, which they may see as a perk. They gain insight into what is happening operationally, which can yield ideas for further research.

The Air Force could consider adopting a model similar to the one at SWPC. Visiting researchers from national labs, service academies, and other government research sites could gain valuable experience and connections by working at AMOS, and the Air Force would in turn benefit from an externally funded labor force that would bolster its R&D mission.

Interviews conducted through in-person conversations at NOAA headquarters in Boulder, Colorado, August 8, 2012 and February 6, 2013.

Keck Provides a Set of General Practices, Remoting Interface Control Procedures

- **Time allocation at Keck determined at the consortium level—each shareholder participates in scheduling.**
  - **Unilateral time allocation by AFRL might lead to tension with AFSPC.**
  - **AFRL should consider steps to ensure transparency in time allocation process.**
- **Keck maintains reliable remote operations by requiring strict technical specifications of remote sites**
  - **Keck's remote ICD available for AMOS's review**
- **RAND judges Keck to be best-of-breed in two areas that are particularly relevant to AMOS:**
  - **Staff continuously assesses downtime metrics in order to ascertain the root of errors and prevent future problems.**
  - **Keck has a very rigid configuration management system that guards against downtime due to unqualified equipment adjustment.**

The W. M. Keck Observatory, located on Mauna Kea, Hawaii, was constructed and operates through the support of a consortium of institutions, including the California Institute of Technology (Caltech), the University of California (UC), and the University of Hawaii (UH). Keck is a purely academic facility created and supported by universities for the purpose of doing academic astronomy research in infrared and optical wavelengths. Keck's two 10-meter primary mirror telescopes are the largest optical infrared telescopes in the world. Keck also houses a suite of powerful instrumentation that includes some of the best available infrared detectors and optical charge-coupled device (CCD) image sensors. Keck has a total staff of about 120. Operational funds come from the universities and NASA. Various partners get a percentage of allocated time. The UC system receives one-third of available time, Caltech another third. As steward of the mountain on which Keck was constructed, UH receives 10 percent of telescope time in exchange for the land lease.

All decisions on day-to-day use of the telescope are made at the level of institutional stakeholders, not at the joint consortium level. Each stakeholder organization has a time allocation committee (TAC) that reviews and ranks proposals and determines how that organization's portion of telescope time will be assigned. Time allocations will be handled differently at AMOS, with AFRL handling allocations for all stakeholders. Keck's more democratic system suggests that AFRL should be on the lookout for possible concerns from stakeholders that AFRL has unilateral control over time allocation at the site. AFRL should take steps to ensure that its time allocation process remains as transparent as possible.[28]

Remote operations capabilities were built into Keck from the start. Keck maintains connections with a series of other sites within a remote observing network. Remote observing from the mainland currently makes up about 50 percent of observations. Remote sites must meet a set of strict technical requirements to qualify, and those who do not meet the requirements must come to Hawaii for observing runs. Keck has developed a set of publicly available remote interface control documents (ICDs) for remote observers that exemplify the rigid rules needed to run reliable, secure remote operations.[29] These might serve as a starting point for AMOS to develop similar technical specifications to enable remote observing.[30]

Keck also serves as an exemplar in two other relevant respects:

- Keck maintains a robust lessons-learned program. Staff continuously track downtime metrics to ascertain the root causes of errors and prevent future problems. This system has led to a decrease in downtime due to mechanical error. Weather currently accounts for about 20 percent of total downtime. AFRL could look to Keck's lessons-learned program as a guide to reduce downtime.
- Project leaders work with a dedicated, point-of-contact support astronomer who is responsible for accurate configuration of the equipment. This ensures that the telescope will not go offline due to adjustments by unqualified external researchers or those seeking to conduct R&D using the observatory telescope and instruments. Even when accommodating visiting PI instruments, some level of staff control over instruments may be advisable at AMOS.

Interviews conducted via phone and e-mail with Keck personnel, June 3–18, 2013.

---

[28] For an example of a planned egalitarian scheduling model, see TMT, Appendix A.

[29] For an example of a planned egalitarian scheduling model, see TMT, Appendix A.

[30] See NASA IRTF, Appendix A, for an example of effective but less-secure remoting procedures that do not require ICDs.

We Also Observed Several Second-Order Lessons from Keck

- **Keck has seen a cultural mismatch between summit workers and sea-level workers.**
  - If AMOS moves toward increased remote observations, AFRL should be mindful of the possibility of a similar mismatch.
  - Keck manages this issue with an informal incentives program.
- **Keck employees involved with low-level outreach, but some staff expressed belief that better public relations could improve financial and cultural support of the observatory.**
- **Pays into Mauna Kea Observatories Support Services for common services like sewage, security, road maintenance.**
  - As ATST comes online, AMOS may want to consider a similar arrangement in order to reduce costs.

Several second-order lessons may also be drawn from Keck's experiences with staffing, outreach, and support services.

Keck takes steps to address cultural mismatches between summit workers and researchers at sea level that have arisen due to differences in lifestyle and work conditions. For example, Keck management openly acknowledges the harsher climate on the summit and the difficult drive for employees up the mountain, and offers incentives for summit workers such as breakfast on days employees drive up the mountain and a four-day workweek for summit workers.[31] If AFRL is concerned about similar cultural issues, AMOS could learn from the steps that Keck takes to ensure that its remote staff are fairly compensated.

Many Keck employees participate in outreach programs intended to garner support for the observatory and its science mission. These programs usually bring outside visitors to the observatory to participate in education programs or send Keck astronomers out to classrooms elsewhere on the island. Undergraduate students have been mentored by Keck scientists, as well. Outreach is conducted on a volunteer basis and organized mainly at the grassroots level. Staff at Keck noted that outreach can be more effective when integrated into the larger mission of an observatory and can pay dividends in popular support. Keck's example plays into a larger theme of the importance of messaging and outreach in building support that AFRL might consider moving forward should it wish to expand its public outreach program.[32]

---

[31] For an example of work scheduling intended to support workers situated in remote or harsh conditions, see ALMA, Appendix A.

[32] For evidence of the success of outreach and messaging programs, see HST/STScI, Appendix A.

All observatories on Mauna Kea pay into communal security services. As more (and larger) tenants arrive on Maui, such shared expense plans might prove useful to bring costs down for such necessary services as water, electricity, data transfer, and security.[33]

Additionally, Keck has engaged a third party to serve as data custodian for the purposes of making data available to a wider audience.[34] The Air Force might wish to observe how this process has worked for Keck as it considers options for making AMOS data available to a wider audience.

Interviews conducted via phone and e-mail with Keck personnel, June 3–18, 2013.

---

[33] Within the next few years, the Maui Space Surveillance Complex will be gaining a new neighbor of comparable size, which might serve as a suitable partner in service sharing. See ATST, Appendix A.

[34] Keck contracts with NEXI to store 60–70 percent of its data, with an ongoing initiative in progress to increase the scope of data storage.

Subaru Facilitates Visiting Researchers that Bring Their Own Instruments

- **The Subaru telescope has four foci, which house a raft of instruments that includes 8 Subaru-owned "facility" instruments and 2 slots for "carry-in" instruments brought by visiting PIs.**
  - **Flexibility for visiting researchers has been built into the machinery itself—a unique value proposition among comparable observatories.**
- **Broadcasting a similar amenability to visiting researchers' instruments could increase the potential customer base for AMOS.**

The Subaru telescope is located on the top of Mauna Kea, where it is used to conduct observations in optical and infrared wavelengths. It is operated by the National Astronomical Observatory of Japan, and research scientists from Japan receive first priority for observing time, with about 30 percent of total time opened to astronomers from the rest of the world. The telescope runs on a classical schedule, with time allocated in increments of full nights and occasional half-night programs. This is partly due to the requirement that at least one member of a research team be physically onsite at the summit during the assigned time slot. Although Subaru only permits a maximum of three people in the control room per research program—the observer, the telescope operator (TO), and support astronomer (SA)—remote observing is mostly not permitted. For observations that make use of stable instruments, the PI or representative of the research team may observe from a remote station at Hilo, but not from any farther afield. A visiting researcher must plan in advance to travel to Hawaii for his or her assigned time slot.

Subaru currently runs eight "facility instruments" and two "PI instruments" at its four foci. The facility instruments remain on-site as permanent assets to the observatory. The two slots for PI instruments allow outside observers to bring their own instruments for use during their observing runs. Subaru conducts a thorough review of all proposals for PI instruments to determine whether or not to devote the time necessary to install a temporary instrument at one of the telescope foci. Of the eight slots available for facility instruments, Subaru has established a review process for new instruments constructed by staff at the telescope itself and in collaboration with other institutes internationally.

By designating dedicated slots for visiting instruments, Subaru has made accommodation of a wide range of visiting researchers a public priority. Rather than treating the exchange of temporary instruments as a nuisance, Subaru management believes that the flexibility of instrument configuration at the site expands the potential customer base to those researchers who might not otherwise apply for telescope time based on extant on-site instruments. Making similar accommodations for potential customers of AMOS might expand the site's value proposition and increase its customer base.[35]

Interviews conducted via e-mail with Subaru personnel, July 17–August 7, 2013.

---

[35] For another strong example of an institutional mindset of welcoming visiting researchers and their instruments, see MIT/LL, Appendix A.

Operations at Pan-STARRS Suggest Survey Data Enables Crowdsourcing, but Requires Special Staffing and Scheduling

- **Survey mission requires specialized scheduling and staffing needs.**
  - **Training staff as dedicated surveyors may lead to greater efficiency rather than relying on cross-trained generalists.**
- **Engineering done *in situ* rather than scheduling dedicated engineering time.**
  - **Consider adopting this practice for the 1.2m.**
- **Engaging a third party data custodian enables dissemination of data to a broader audience.**
  - **Recognize that a dedicated data custodian may lead to more effective information sharing.**

Pan-STARRS is a wide-field imaging facility developed at the University of Hawaii's (UH's) Institute for Astronomy. Two telescopes, Pan-STARRS 1 (PS1) and Pan-STARRS 2 (PS2) will be in service. The prototype single-mirror telescope PS1 is operational on Maui, with a scientific research program currently underway by a consortium of research organizations. PS1 was built to conduct wide-field surveys of large parts of the night sky. A major goal of Pan-STARRS has been to discover and characterize near-earth objects (NEOs), both asteroids and comets, that might pose a danger to the planet. The Air Force supplied the initial funds to build PS1 but not to operate it. The Air Force also supplied money to build PS2, expected to be operational in 2014, but that funding was cut in 2011. UH is responsible for finishing construction of PS2.

Operational funds for PS1 were initially provided solely from academic institutions, although NASA later supplied some funds, as did the Air Force. The budget is $5.5 million for PS1 and PS2 combined, which does not include NEO software upgrades and operations. There are 18 full-time employees, most of whom are technical staff. Operations are dedicated mostly to astronomical research, and outside agencies are allowed to use the facility in exchange for use of data for other astronomical research purposes.

The operational methods in place at observatories that engage in full-sky surveys, such as PS1, are different than those at observatories that cater to traditional, narrow-field observations. This includes differences in staffing and scheduling. For example, at PS1 all surveys are conducted by specialized astronomers on staff, as opposed to more generalist astronomers at traditional, narrow-field observatories who serve as on-site advisors for visiting researchers. In addition, survey data can be used for multiple purposes. Rather than scheduling breaks in the observing schedule for engineering and instrument adjustment, as is the practice at many narrow-field observatories, it is possible to conduct engineering *in situ* using telescope data taken during wide-field surveys. Images of an object taken one night, under specific atmospheric and technical conditions, can be compared to an image of the same object taken on another night under different conditions, and the difference used to calibrate and adjust instrumentation. Engineering is therefore built into operations. AFRL should consider adopting this practice for the 1.2-meter telescope that is slated to conduct surveys at AMOS so as to maximize the amount of telescope time used for observations.

Processed data from PS1 are currently initially made available to researchers in the consortium, though PS1 has plans to make the data more widely accessible. Pan-STARRS has made plans for STScI to serve as its data custodian in the near future, for the purposes of making the rich data collected from the surveys available to a wider audience.[36] Because STScI will be a Pan-STARRS partner in the future, the costs of data storage will be transferred away from the operations budget where it currently constitutes a major cost. As demonstrated by multiple case studies of observatories presented here, making data more widely available to additional customers may extend the impact of an observatory.[37] By engaging a dedicated data custodian—either internally or by engaging a third party as Pan-STARRS plans to do—to manage storage and sharing, the data collected at AMOS may serve an extended audience of researchers.

Pan-STARRS is a neighbor of AMOS on Haleakala, and could serve as a partner in certain information sharing capacities. Weather monitoring could be consolidated and supported by all major observatories on the mountain. Weather data could be improved for all observatories on Maui by setting up one standard weather station to serve all sites in the region with comprehensive weather data. AMOS might consider developing such partnerships with its major neighbors, including Pan-STARRS and ATST.

Interviews conducted via phone and e-mail with Pan-STARRS personnel, June 7–July 12, 2013.

[36] For more information about its data storage and access programs, see STScI/HST, Appendix A.

[37] For examples of observatories that have prioritized data storage and sharing and the impact of such programs on knowledge production, see STScI/HST, UKIRT, Keck, Gemini, ALMA, and IRTF, Appendix A.

Agile Hardware Swaps Enable Smaller Time Allocation Increments at NASA IRTF

- IRTF is able to allocate time into half-hour increments due to agility of hardware swaps.
  - This suggests one way that AMOS could improve efficiency by reducing the time it takes to switch between customers.
- Instrument control has been fully remoted, and some shorter programs can be executed by software.
- IRTF incorporates a human into the loop in order to provide critical thinking and social comprehension to the scheduling process.
  - This suggests that it may not be possible to fully automate scheduling in a complex system

The NASA Infrared Telescope Facility (IRTF) consists of a 3-meter telescope funded entirely from the NASA Planetary Astronomy program. Its primary mission consists of mission support and full system observations for NASA—in general, conducting observations that are required for NASA missions. About 50 percent of the time at IRTF is dedicated to solar system observations and the other 50 percent of the time for non–solar system observations. The National Science Foundation (NSF) provides some funding to support new instruments in service of non–solar system observers, and IRTF also supports PIs conducting basic solar system research that may or may not be related to NASA spacecraft missions.

Many IRTF observations are conducted remotely by independent researchers all over the world. Because they may control the instruments from their home institutions, PIs do not have to make travel arrangements. Observing runs may be scheduled with relatively short notice, total observing time for an individual PI may be broken into smaller blocks if conditions require, and smaller observation increments for different PIs may be scheduled in quick succession. Anyone with an Internet connection may observe, with the assistance of an on-site TO to slew the telescope and make larger adjustments. Shorter programs may be executed by software. Only two staff members are required to be in the dome at night. Support astronomers are only called in if an anomaly occurs. The tradeoff to this relatively open-access model of remote observing is in a lower level of security and the possibility of downtime due to Internet connectivity or server failure.

This flexibility is also enabled by instruments that can be changed at night. Programs that require different instruments can be scheduled on the same night, or spread out over several nights, depending on scheduling needs and external conditions. Having instruments that can be switched on the fly allows tighter scheduling on the classical schedule model, but may be significantly more expensive than traditional instruments that require more time to swap out.[38] Tight scheduling is also facilitated by the presence of a single customer, NASA, rather than the consortium of funders that supports observatories of comparable size. AMOS might be able to improve efficiency by considering investment in more-agile equipment swaps, facilitated by a strong configuration control system that would reduce the downtime needed for calibration across all systems.

Interviews conducted via phone with NASA IRTF personnel, June 26, 2013.

[38] We have seen indications that investments in either agile instrument swapping capability or improved scheduling software can increase efficiency in telescope use. See Gemini North, Appendix A, as an example of an observatory whose instruments are set during the day and static at night but that has achieved the same level of incremental scheduling as IRTF by using a queue schedule.

Gemini North Suggests Pros and Cons of Queue vs. Classical Scheduling

- **Queue scheduling allows for maximum efficiency in telescope use by enabling the following:**
  - **Matching of existing conditions (maintenance, instrument status, etc.) to project requirements**
  - **Projects that are higher ranked have a better chance of getting data; higher prestige for observatory**
- **Requirements for queue scheduling:**
  - **Robust technical infrastructure supporting remote observing**
  - **More information considered during proposal review process**
  - **Reliable weather data**
- **Gemini is the exemplar of queue scheduling, and would be a good place to learn more about it.**

The Gemini telescopes use an 8.5-meter primary mirror to observe in infrared and visible wavelengths. Gemini North is the northern hemisphere component of a dual-telescope observatory. Gemini South is an identical telescope in Chile, which means that the two telescopes together provide coverage for the entire sky. Gemini is run by an international consortium of six partner countries and the University of Hawaii, each of which receives a percentage of total telescope time roughly proportional to their capital contributions.

Gemini runs its scheduling on the queue model. Queue schedules are designed to maximize telescope time by ranking observing programs based not only on scientific merit and target availability but also on required conditions for observing. Unlike classical scheduling, in which all PIs receive a specific night or nights for observing regardless of weather, technical, or seeing conditions, queue scheduling enables those projects that are more highly ranked to be bumped up in the queue during prime conditions and accounts for those proposals that can be carried out during conditions that might prohibit other projects from getting data.[39] With queue scheduling,

---

[39] It is important to note, however, that studies of publications produced by higher-ranked proposals do not necessarily lead to greater numbers of citations, a metric often used to gauge the impact of a scholarly publication. See Dennis Crabtree and Elizabeth Bryson, "Observatory Publications and Citations." Library and Information Services in Astronomy IV (LISA IV), Emerging and Preserving: Providing Astronomical Information in the Digital Age. Proceedings of a conference held at Charles University, Prague, Czech Republic, July 2–5, 2002. Edited by Brenda G. Corbin, Elizabeth P. Bryson, and Marek Wolf. Washington, DC: U. S. Naval Observatory, 2003, p. 204–206.

it is less likely that an observer will be "clouded out" or otherwise miss the opportunity to collect data—although the queue system is not foolproof in this respect. Parameters such as cloud cover, moonlight, and water vapor can be predicted ahead of time, but seeing must be measured in real-time so that programs may be picked up or dropped from the queue lineup depending on changing conditions.

Queue scheduling is largely enabled by a robust remoting infrastructure that facilitates observers from far afield to conduct observations without coming to Hawaii. Support astronomers on staff at Gemini work with individual PIs in advance to coordinate an observation plan, and data are delivered to the PI in real time. Unlike IRTF, all Gemini instruments must be swapped and set up during the day, which adds another ingredient to the scheduling mix. A larger suite of information must be gathered from the PIs during the proposal process to determine how cloud cover, moonlight, water vapor, and seeing might affect their research program, so that the queue may be rearranged based on these factors. By 2015 or 2016, Gemini hopes to be operating fully remotely from Hilo, with nobody in the dome at night. Currently, the adaptive optics system is too temperamental to be managed remotely, as it requires regular adjustment.

If AFRL becomes interested in adjusting its scheduling protocol, it might look to Gemini North for an example of the pros and cons of classical and queue scheduling. The in-house proprietary scheduler developed by Gemini might also serve as a potential scheduling software model.

Interviews conducted via phone with Gemini North personnel, June 25, 2013.

TMT Suggests Methods of Egalitarian Scheduling and Controlled Flexible Configuration

- **Each partner of the TMT consortium will be responsible for its own time allocation.**
  - Partners allowed to schedule using classical or queue, and each partner may manage its own mini-queue
  - This model may be useful as an example of ways to make sure AMOS partners have adequate representation in the time allocation process
- **TMT approaches flexible configuration with caution**
  - Instruments must be fully tested and calibrated before use, but this requires downtime
  - Necessary to establish formal specifications for new instrument installation and testing
  - AFRL should consider developing a set of specifications to be codified and internally shared, so as to avoid conflict over configuration-related down time

The Thirty Meter Telescope (TMT) is currently under development, with construction scheduled to begin off the summit of Mauna Kea in April 2014 and first light expected at the end of 2022. The telescope will observe in wavelengths ranging from near ultraviolet to visible, and will incorporate a wide-field optical spectrograph for observations in the visible spectrum. Project managers expect that the unprecedented size of the aperture and its AO systems will make TMT one of the largest and most efficient ground-based telescopes in the world upon completion. Possibly due to the AO system on-site, TMT will be remotely accessible by PIs, but a TO and support astronomer will be on-site at all times. The project is supported by an international consortium including the UC system, Caltech, the Association of Canadian Universities (ACURA), and the national astronomical observatories of Japan, India, and China.

Each of these partners will be allotted a percentage of total observing time, and each partner will be responsible for conducting time allocation within that allotted percentage. Partner institutions may choose to follow a classical or queue scheduling model, and those that adopt a queue system may run their own autonomous mini-queues. This planned method of egalitarian scheduling serves as an example of one way to be sure that all sponsoring institutions exercise some amount of agency in the time allocation process, which may lead to less potential conflict over telescope time.[40] AFRL might consider watching how TMT's scheduling initiative pans out, especially if conflict arises regarding AFRL's sole discretion for time allocation.

[40] For a similar approach to egalitarian time allocation see Keck, Appendix A.

TMT management acknowledges the importance of allowing new instrument development, but approaches changes to telescope configuration with caution. While instruments should ideally be fully calibrated and tested before being brought into regular use, such a testing period requires downtime that could otherwise be used for observations—an eventuality that can cause frustration among astronomers. However, instruments that have not been thoroughly tested before use may succumb to failure, which also frustrates astronomers who have invested time, effort, and money into building and executing a research plan. TMT management plans to draw up a set of formal specifications to control new instrument installation so as to avoid some of these possible conflicts. Currently, TMT plans to add a new facility instrument every four years. When considering flexible configuration at AMOS, AFRL should think about developing a similar set of specifications to be shared among its partner institutions, so that all interested parties may be made aware of the necessity of downtime for instrument testing and the details of such procedures.

Interview conducted via phone with TMT personnel, July 19, 2013.

MIT/Lincoln Laboratory Provides Guidelines for Running an Effective R&D Enterprise

- Installation of new equipment left behind by visiting researchers seen as an asset, not a liability.
  - AMOS always needs to be on the lookout for how visiting researchers can benefit AMOS
  - Should be aware of benefits from day one, before the visiting experimenter arrives
- Successful operation of an R&D facility for 20 years with stable funding enabled by:
  - Everyone from technicians to managers can accurately discuss technical advantages
  - Always tie work back to the sponsor
  - Can make and test changes the same night

The Aerospace Division at MIT Lincoln Laboratory (MIT/LL) oversees several sites relevant to the challenges of running operational and R&D facilities. The Space Situational Awareness Group runs the Lincoln Space Surveillance Complex, which consists of three radars used for tracking space objects. The mission of the Complex is therefore related to the telescopes at AMOS—though still very different in their operations and products. MIT/LL also fields many secondary customers that are sensitive—another attribute in common with AMOS.

The MIT/LL Space Control Systems group developed the ground-based Space Surveillance Telescope (SST), which also provides a model for an effective R&D shop. Funded by DARPA, SST is very much a pure laboratory—the data collected through SST do not go directly into the Space Surveillance Network. Instead, they are fed back to DARPA for analysis to provide feedback to the Air Force. Because they work at a true lab, SST researchers have more leeway to try new technologies and procedures, with the expectation of a certain rate of failure. Because of this mindset, SST is extremely open to allowing visiting researchers to bring their own instruments to the site, and is dedicated to finding new customers for instruments left behind. Configuration management is handled by technicians who are experts at operating all equipment and all programs, a requirement that is essential to the institutional philosophy at SST. Software and technical developers can make and test changes the same night, and the same small core group of developers handles the configuration and testing. The close knowledge of the on-site technicians reduces the need for additional training time, which streamlines procedures and allows the rest of the team to devote time to building the domain knowledge base.

Of the many lessons learned from the MIT/LL installations that we examined, two main recommendations stand out with regard to how best to run an effective, well-funded R&D enterprise. First, MIT/LL welcomes visiting researchers who bring their own instruments to the site. Rather than seeing these instruments as incurring extra costs in time and money to install and run, the management at MIT/LL sees these instruments as value added in terms of a broader potential customer base—both in attracting researchers who want to bring their own instruments and researchers who might wish to use said instruments when they are left behind on-site. Management uses newly acquired instruments to target potential researchers who might be interested in data collected using the new setup. This mindset of extra value gained by fielding visiting researchers' instruments may be one way that AMOS might broaden its value stream and attract additional customers.

Secondly, the higher-level management of MIT/LL has emphasized the importance of a strong, institution-wide focus on MIT/LL's mission and understanding of its value proposition. All employees, not only managers, are aware of the technical advantages of the site and can discuss them accurately, with the understanding of how these advantages work in service of the customer. This has been worked into the culture of the institution through mentoring relationships and regular communication across different staff groups. While precipitating a cultural shift may be daunting, simple programs of staff education and communication may go a long way toward facilitating such an institution-wide awareness at AMOS.

Interviews conducted via phone and e-mail with MIT/LL personnel, June 27–July 18, 2013.

MIT Radar Operations Cannot Be Directly Compared to AMOS Optical Operations

- **Operationally, radar is not directly comparable to optical for several reasons:**
  - Physics of radar allows for less latency
  - Radar data easily integrated into nation's SSA infrastructure
- **MIT/LL radars run by FFRDC employees, not contractors.**
- **MIT/LL not comparable to AMOS because the presence of a single funder (21st SW) provides site with two key advantages:**
  - Tighter value stream for radar data
  - Absence of visiting experiments means stable hardware configuration, less downtime

Although several of these lessons are of interest to AMOS, MIT/LL radar operations in particular are not directly commensurate with operations at AMOS. On a basic level, the physics of radar observation is different from the optical observation conducted at AMOS and other optical sites. Because radar does not require the same weather, darkness, and seeing conditions that optical observations require, radar data can be collected more regularly and with less latency. Additionally, the United States' space surveillance infrastructure was initially built to use radar data. Given that other facilities do not fully support optical data, the value stream of radar data collected by sites such as the MIT/LL radars to the nation's SSA is more direct than optical data. Another major difference between MIT/LL radars and AMOS is its employee base: AMOS is currently run by staff provided by a contract with Boeing, Schafer, and Wolf Creek, whereas MIT/LL staff are FFRDC employees. The MIT/LL radars are funded by a single entity, the 21st Space Wing, which means that the data being collected by these facilities flows directly to a single customer, thus tightening the site's value stream. The presence of a single customer means that MIT/LL does not need to field additional customers. As a result, the hardware configuration does not need to be flexible enough to facilitate visiting researchers. At the radar installations, stable configuration means less overall downtime for the radars, and a steadier rate of data collection overall. These three main considerations make comparison of operations at AMOS to MIT/LL's radar operations imperfect, but the lessons learned on running an efficient R&D shop, as presented in the previous slide, may still prove useful for streamlining R&D at AMOS.

Interviews conducted via phone and e-mail with MIT/LL personnel, 6/28/2013-7/18/2013.

LCOGT Provides Lessons on How to Run a Remote Network, Establish a Pricing Plan

- Provides insight on the ingredients needed to run an automated telescope network:
  - Standardized and uniform parts
  - Identical telescopes within each cohort
  - Extensive telemetry for problem detection
  - Uniform mount control software across network
- LCOGT provides a minimum baseline for what it costs to run an automated telescope network:
  - $300/hr. for 1m
  - $600/hr. for 2m

The Las Cumbres Observatory Global Telescope Network (LCOGT) was built upon the private purchase of two 2-meter class telescopes that were previously used to allow schoolchildren the chance to conduct observations. With its initial two telescopes situated in Hawaii and Australia, the LCOGT observing model is based on geographical distribution – with enough telescopes far enough apart, observers will always be able to have access to a telescope in night. Although some school programs were grandfathered into LCOGT, 80 to 90 percent of total telescope time is devoted to projects by academic researchers who pay a set fee for use of the telescope. The network has grown to consist of ten 1-meter telescopes, two 2-meter telescopes, and two half-meter telescopes distributed at sites around the world. Astronomers conducting time domain studies represent the ideal customer: an observer who will take advantage of the widespread network. Each cohort of telescopes is nearly identical, running the same hardware and instrumentation. A single piece of software runs all the sites as a unified network.

In order to run a successful automated network of telescopes, several technical specifications must be in place. Parts must be standardized and uniform across all telescopes of the same class. Extensive, reliable telemetry must be consistently gathered from detectors and actuators situated at all points of possible failure, even those that may be largely dormant. Finally, mount control software must be uniform across the entire network—all telescopes of all classes must be controlled using a standard set of software. AFRL should keep in mind that standardizing software across mounts may increase efficiency at a smaller geographical scale, as well.

LCOGT has developed a pricing system that constitutes its primary funding source, as well as democratizes use of the telescope. The current pricing is set at $300 per hour for use of one of the 1 meter telescopes, and $600 per hour for use of one of the 2 meter telescopes. Price breaks are made available for high-volume requests. Because the network's schedule is not yet oversubscribed, any paying customer may apply for time and the scheduling takes place via software, with occasional intervention by human schedulers for particular celestial events, such as supernovae. AFRL might consider making up a pricing plan in order to draw customers to AMOS.

Interviews conducted via phone with LCOGT personnel, June 27, 2013.

ATST May Be a Potential Future Partner in Sharing Infrastructure, Labor Costs

- **AMOS is about to get a neighbor of comparable size and complexity. ATST represents an excellent possibility for a partnership of mutual benefit with AMOS.**
  - Will share mirror coating facilities
  - Potential partner on infrastructure costs
  - Potential customer for MHPCC
  - Potential partner in labor resources
    - AMOS might consider creating an AFRL-funded fellowship for a graduate student to work at ATST

The Advanced Technology Solar Telescope (ATST) is currently under construction on Maui. ATST is part of the National Solar Observatory, which is operated by the Association of Universities for Research in Astronomy under a cooperative agreement with the National Science Foundation. The main mission of the National Solar Observatory is basic science, specifically a focus on the astronomy of the sun and how the sun fits into the universe through the study of solar magnetism, especially variations in solar output, as well as energetic phenomena driven by magnetism and the solar atmosphere. ATST will be the world's largest solar physics observatory. It will be a 4-meter class facility with a set of instruments that can collect data in visible and infrared wavelengths.

There will be two parts to operations at ATST. Maui staff will operate the telescope, conduct observations, and collect data. The headquarters in Boulder, Colorado, will house additional staff and science operations. The data center will be in Boulder. Costs for the Maui component will be approximately $10 million per year, and those in Boulder will be about $8 million, for a total operating budget of about $18 million in fiscal year 2019.

At the current National Solar Observatory site in Sunspot, New Mexico, project leaders are allowed to do hands-on observing and R&D on the telescope—there is no configuration management. At the Maui site, an emphasis on operations is expected, though some room for R&D will be maintained. At ATST, instrumentation specialists and support astronomers will be in charge of operating the equipment and will support project leaders. This "service mode" approach is different from the model used in New Mexico. In the future, ATST management will need to determine how best to balance R&D and operations. AFRL may want to check with them again in five years to see how they have fared.

The plan is for ATST to make public all data collected at the observatory. Data from ATST will be relevant to such issues as the electrical grid, infrastructure, and climate change, among other concerns. A plan for data sharing is being developed prior to the observatory's completion. The Air Force may want to consult with ATST about their data-sharing procedures when they are in effect.

ATST may experience cultural rifts between island/mainland staff and observers. If this problem occurs, the Air Force should note how ATST addresses it, or may want to offer advice on what to expect from this kind of divide, and how best to avoid it.

As a neighbor on the mountain, ATST may present opportunities for partnership with the Air Force to save money on infrastructure needs, including water, power, weather information, data transport, sewage, and security. AFRL and ATST might also consider sharing labor resources in the form of an AFRL-funded fellowship for a graduate student to work at ATST.[41]

Interviews conducted via phone with ATST personnel, June 18, 2013.

[41] For another relevant example of labor sharing with appropriate and local institutions see SWPC, Appendix A.

UKIRT Provides an Exemplar of Efficient Survey Operations

- **UKIRT runs a highly automated scheduling system, supported by proprietary scheduling software.**
  - **If AMOS seeks to streamline its queue scheduling, UKIRT's model might serve as a model.**
- **The number of publications produced from UKIRT data has risen exponentially following an effort to refocus its mission to a particular niche (wide-field infrared observations) and making data available to entire UK community following a yearlong embargo.**
  - **A streamlined, targeted value proposition can bring in customers even when services are reduced.**
  - **Making data more widely available, even internally, can lead to greater impact and production levels**

The United Kingdom Infrared Telescope (UKIRT) is a 4-meter telescope that observes in infrared wavelengths. It is funded entirely by the United Kingdom. In recent years, budget cuts have forced UKIRT to operate in what personnel call "minimalist mode." This mode of operations runs on a skeleton crew of two staff scientists and three TOs, and has forced the observatory to go fully remote. Staff members control the telescope from a site in Hilo. UKIRT is currently looking for a buyer for the telescope, but in the interim continues to run the telescope on a much more compact scale with a narrower mission scope.

UKIRT runs one of the most automated scheduling systems we have seen. UKIRT built its own proprietary software to schedule observing time based on a queue model. While UKIRT's scheduling system still includes a person in the loop for the purposes of updating current weather conditions, reprioritizing projects as necessary, and other on-the-fly considerations, it is otherwise a fully automated queue scheduler. The system produces the top-ranked project that fits current conditions, without the input of a human scheduler. If AMOS wishes to streamline its queue scheduling protocol, UKIRT's system might serve as an efficient model.

Due to budget cuts and the resulting reduction in staff, UKIRT lost some of its science capability in the form of instruments that have been taken offline. UKIRT has focused instead on a specific niche catered toward a smaller field of projects. Operating mostly in wide-field mode, UKIRT has managed to see an exponential increase in the number of publications produced using data collected on-site. This is considered to be a result enabled by a simultaneous change in UKIRT's data access policy. UKIRT embargoes data for a year after collection, but anyone who is a part of the UKIRT consortium can get immediate access to the data during the embargo. After the embargo, anyone in the UK may access the data. As a result, the number of publications produced since this policy change has increased by about 200 per year. This drastic increase suggests that AFRL might want to consider expanding access to AMOS data, even within constrained parameters, in order to increase the impact of the data.[42]

UKIRT is remarkably efficient with engineering, reducing the time devoted to maintenance and upkeep from 30 percent to 2 percent within six years. This was made possible by the low level of failure in wide-field mode. In addition to the *in situ* engineering feasible in this observation mode, AFRL should keep in mind that the 1.2-meter telescope at AMOS may require less time for engineering than the 1.6- and 3.6-meter telescopes.[43]

Interviews conducted via phone and e-mail with UKIRT personnel, July 18–19, 2013.

---

[42] For data on the impact of stored and shared data, see STScI/HST, Appendix A.

[43] For more information on *in situ* engineering in the wide field, see Pan-STARRS, Appendix A.

Kwajalein Missile Range Promoted Investment by Incentivizing Early Adopters

- **Before 1994, DoD provided test services free of charge. Customers requested all available data regardless of actual need. To reduce costs, DoD reduced funding provided to test range customers to 40% of baseline.**
- **To recoup the remaining 60% of baseline, Kwajalein implemented a business plan:**
  - **Basing fees on an "a la carte" model forces customers to think carefully about what they want, and the value of the product.**
  - **Kwajelein attracted early investment by incentivizing pilot adopters.**
- **AMOS should consider this approach as one way to attract customers**
  - **Payback scheme**
  - **Early investment at a lower rate**

The missile test range located on Kwajalein Atoll, currently known as the Ronald Reagan Ballistic Missile Defense Test Site, is one installation of the Major Range and Test Facility Base (MRTFB). The MRTFB and Nellis Air Force Base outside of Nevada facilitate large bomb testing. These sites are seen as national assets and infrastructure—they are not always in use, but the doors must be kept open so that they are ready when the need arises. Before 1994, the DoD provided services at these test sites free of cost. Because they were free, customers expected to be given all data collected regardless of actual need, and costs skyrocketed for the DoD. The bill for operations continued to rise steadily until the DoD changed its policy so that it would subsidize only 40 percent, with the remaining 60 percent paid by customers.

In order to make sure that the 60 percent of baseline costs were covered and Kwajalein could remain open for business, a business plan was developed and implemented based on incentives and a fee schedule. The model developed and successfully implemented at Kwajalein suggests several ways to ensure customer commitment and keep operational costs within reason.

First, a fee schedule must be established for data produced. At Kwajalein, customers were charged for data products "a la carte," which made customers think carefully about which data would be most valuable to them.

Second, the best way to attract early money is to incentivize. Customers may commit early if they are promised something special. Some options include a payback scheme and the option to invest early for a lower rate.

Third, once a business plan is drawn up, those at the helm have to engage in a "roadshow." To sell the program, one must accurately explain site capabilities, and tie it back into customer mission every single day.

This subsidy and incentive structure could be applied to AMOS's approach to soliciting and accommodating new customers. By offering early investment in exchange for a locked-in lower rate as well as a payback scheme, AFRL could establish a committed customer base for AMOS.

Interviews conducted via phone with former Kwajalein Missile Range personnel, July 18, 2013.

Kwaj Cannot Be Directly Compared to AMOS

- **Kwaj cannot be considered comparable to AMOS in terms of support structures:**
  - **The military installation at Kwaj supports a self-contained, self-sustained community: all resources to support operations and staff brought in, including hospitals, schools, retail, etc.**
  - **These resources exist independently of the Air Force on Maui.**
  - **Kwaj runs on an a la carte model, whereas AMOS runs on a subscription model under the current MOU**

Although the installation at Kwajalein provides an excellent example of a business plan drawn up to address a changing funding landscape, in certain key respects Kwaj cannot be directly compared to AMOS. In particular, support services and infrastructure that are already in place on Maui independent of the Air Force must be brought in to Kwaj by the DoD. Everything from hospitals, schools, and homes to retail establishments did not exist on Kwajalein prior to the military installation.

Additionally, AMOS currently runs on a subscription model based on the standing MOU with its main customers, rather than the a la carte model in place at Kwaj. Should AFRL consider charging for data on a similar a la carte model in the future, any unused data should not be discarded, but stored for use by other potential customers.[44]

Interviews conducted via phone with former Kwajalein Missile Range personnel, July 18, 2013.

---

[44] For evidence of the value of storing and sharing data see UKIRT and STScI/HST, Appendix A.

NOAO Provides Lessons on Land Use in Culturally Sensitive Areas

- Land use on Kitt Peak has changed as the number of NOAO subletters has increased.
  - Similar trend on Maui
  - Requires greater management of the relationship between landowner and tenants, sub-tenants
  - Important to establish processes to manage the relationship between other tenants and AFRL
- Necessary that observatory tenants understand the political and cultural ramifications of using sensitive land.
  - Symptomatic of astronomer culture to forget terms of land use
  - Important to recognize that not all nations receive the same compensation for lease arrangements

The National Optical Astronomy Observatory (NOAO) runs several observatory complexes for the NSF, including the observatory located at the summit of Kitt Peak in Arizona. The land is owned and leased by the Tohono O'odham Nation, and the lease with the NSF for the Kitt Peak National Observatory has been in effect for over 50 years. Increasingly, NSF funds and support are being applied disproportionately towards the Cerro-Tololo Interamerican Observatory (CTIO) in Chile, which is seen by some at the NSF as a newer, more state-of-the-art facility compared to the more venerable Kitt Peak.

As NSF support has shifted away from Kitt Peak, scientific residency on the mountain has changed. Although NOAO is the primary tenant on the mountain, in recent years an increasing number of sub-tenant groups have set up shop and now outnumber NOAO-run telescopes. Not all sub-tenants are aware of the full terms of the lease and may violate certain terms because of their effective distance from the legal agreements in place. As suggested to us by staff at NOAO, it is often symptomatic of astronomer culture to forget the agreements for land use and assume that all astronomical practices are benign. AFRL should look to NOAO and Kitt Peak for an example of the necessity for managing sub-tenants and for respecting the initial legal agreements between the land user and those who either own the land or hold the land in trust. Part of this respect requires a greater understanding among all tenants of the political and cultural impact of constructing and operating observatories on ecologically and culturally sensitive lands.

Interviews conducted via phone with NOAO personnel, July 18, 2013.

STScI Has Achieved Unmatched Impact Through Effective Outreach and Data Accessibility

- In addition to staff in the Director's office who interface with government officials, a dedicated division handles publicity for HST.
  - HST's value proposition is well known to both the public and to lawmakers.
  - AMOS should devote more resources to outreach so that internal customers and the US government are aware of its value to the nation.
- STScI supports a data archive system that enables an unparalleled level of accessibility.
  - After one year of embargo, anyone across the world may request data from the archive system.
  - Database searchable by parameter: target, instrument, etc.

The Space Telescope Science Institute (STScI), which supports the Hubble Space Telescope (HST), is an exemplar of outreach and external messaging. STScI maintains a staff division dedicated to publicizing Hubble's accomplishments and to keeping its contribution to astronomy in the public mind—an important task when it comes to garnering legislative support. To an extent arguably unmatched in modern astronomy, the value proposition and scientific and cultural impact of HST is well known to the American public and lawmakers.

While the exact approach to outreach embraced by STScI cannot be directly applied to facilities with sensitive customers, AMOS could learn from the way that STScI prioritizes external outreach. By keeping those who can affect funding consistently in the know about a site's value and contributions, a staff dedicated to outreach can directly impact the consistency of support among stakeholders. AMOS could adopt a similar plan, constrained to customers and decisionmakers within the U.S. government.

Hubble provides an excellent example of how an effective data storage and access system can greatly expand the impact of data collected at a single site. The Space Telescope Science Institute runs a searchable database accessible to nearly everyone in the world, providing open access to data following a yearlong embargo during which only the PI who requested and was awarded telescope time can use the data. STScI has collected data on publication rates that illustrate the great extent to which an open access data storage system can increase the impact of data collected by a single PI.[45] AFRL might keep these numbers in mind when considering the possibility of implementing a data storage and sharing plan.

Interviews conducted via phone and e-mail with STScI personnel, August 21–23, 2013.

---

[45] UKIRT has also collected data on the increase of publications based on archived data. See UKIRT, Appendix A.

ALMA Runs a Fully Remoted Operation with Unique Staffing Structure

- At 5000 meters, ALMA is one of the most remote observatories in the world, with only security staff regularly on-site once it becomes fully operational
  - Support staff circulate through remote control site at 3000 meter level, with regular intervals of staffing.
  - All PIs work from home institutions, with a primary liaison at each regional center who communicates project agenda to operators at 3000 meter level
  - Extreme conditions at ALMA make it a useful model of personnel procedures for remoted facilities
- Dynamic queue scheduling enabled by a multi-tiered TAC and scheduling software currently in development
  - Software will be overseen by a manager, but otherwise will change programs and instruments autonomously based on weather, seeing conditions
  - Autonomous software, when developed, may provide an exemplar for more efficient queue scheduling

The Atacama Large Millimeter/Submillimeter Array (ALMA) is currently under construction at 5,000 meters above sea level on the Atacama Desert plateau in Chile. Although the full complement of 66 antennas has yet to be completed, the site has been operational since 2012 following the successful use of the initial set of 16 antennas. ALMA is unique not only in its remoteness and size, but also in the type of data collected in high-frequency wavelengths. At these wavelengths, ALMA researchers are able to detect thermal emissions and conduct chemical analysis of regions of star formation, the atmospheres of planets forming around new stars, and the rate of star formation in the early universe. Much of this information is absorbed by water in Earth's atmosphere, so Atacama's status as the highest, driest place on Earth makes it ideal for this kind of observation.

In order to facilitate data collection at such a remote site, ALMA was designed to be operated fully remotely. All support staff work from a remote control site at 3,000 meters, with staff circulating in intervals between the remote control site and ALMA headquarters in Santiago. No PIs visit the remote control site, or even headquarters—they remain at their home institutions and interact with a support scientist at one of the regional centers located in the United States, Germany, and Japan, who then communicate the PI's science agenda to staff at the remote control site. The only staff that will be present at the 5,000-meter level of the observatory once construction is complete will be security staff on-site to prevent theft. Except for energy and data lines, no other services or infrastructure will be established. To some extent, this high level of remoting is made possible by the nature of radio observing—although conditions must be right for the desired wavelengths to reach the antennas, the antennas themselves are hardier than optical telescopes. It is also facilitated by a labor schedule that keeps on-site staff from being too deeply impacted by the harsh climate and isolation. AMOS should look to ALMA for an example of personnel scheduling should it decide to move towards remote operations.[46]

ALMA is currently developing a new kind of scheduling software that will enable dynamic queue scheduling. Unlike regular queue scheduling, which can be completed mostly in advance for programs that are not contingent on weather conditions, such as those at lower frequencies, dynamic queue scheduling allows for changes to the schedule on very short notice by breaking up larger projects into smaller subunits ranging from half an hour to one or two hours. Once each project is started and all calibrators are in place, shorter projects will be run to completion. For longer projects that take place over many hours, the variability of weather is taken into account—if the weather changes, ALMA's scheduling software will change projects and delay the remaining subunits until prime conditions return and run programs more amenable to extant conditions. When this software has been fully developed and tested, it may serve as a valuable model for AMOS should AFRL wish to further streamline its queue scheduling protocol.

Interviews conducted via phone with ALMA personnel, September 6, 2013.

[46] For more on the necessity of accommodating and incentivizing remote staff, see Keck, Appendix A.

Appendix B. University of Hawaii Telescope Time Allotments

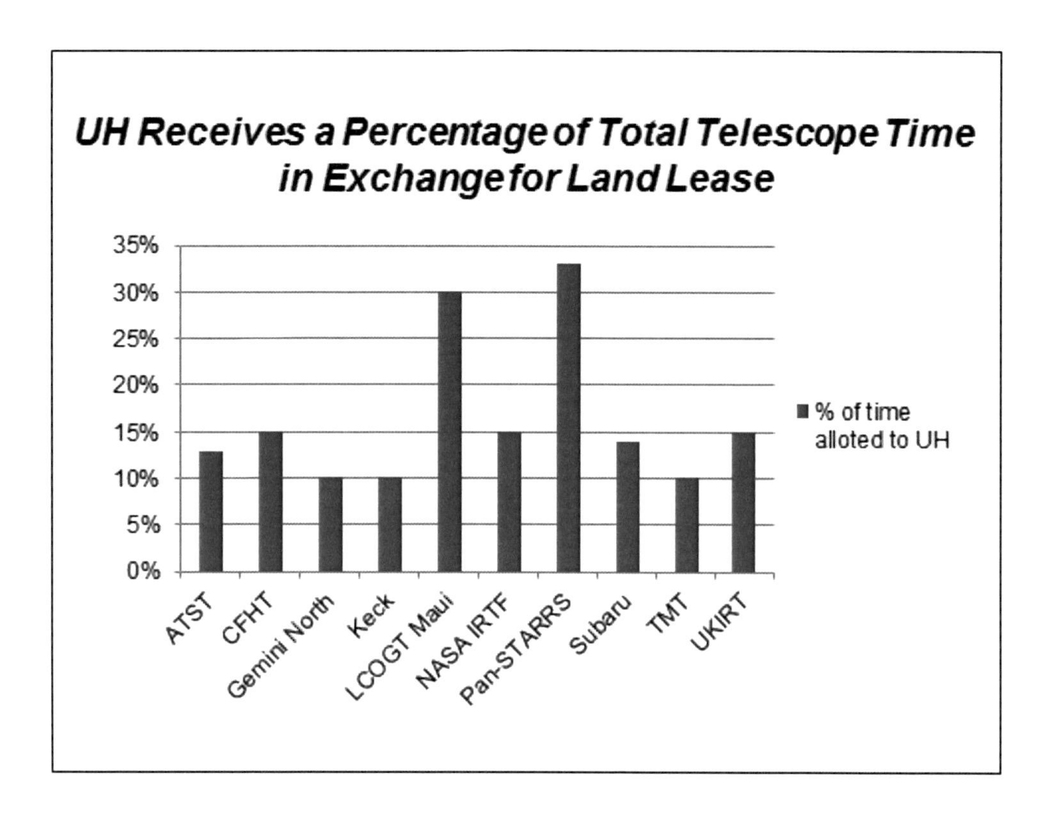

In this appendix, we describe how observing time at observatories located in Hawaii is allocated to the University of Hawaii as part of land lease agreements, and how this percentage of time varies between observatories.

As we progressed through interviews with observatories based in Hawaii, we noticed a pattern in time allocation: A percentage of total observing time at each observatory is dedicated to research conducted by scientists affiliated with the University of Hawaii (UH). The standard amount appeared to be about 10–15 percent of total observing time, with LCOGT and Pan-STARRS as major outliers with 30 percent or more of total observing time dedicated to UH. The land lease agreements that these observatories hold with the State of Hawaii requires this time exchange, and varies from lease to lease.

We present this information only because—to our knowledge—it has not been published elsewhere before.

Appendix C. Organizations That Contributed to This Research

We Conducted Interviews with Representative from Several Institutions

W.M. Keck Observatory
Panoramic Survey Telescope and Rapid Response System (Pan-STARRS)
Kwajalein Missile Range
Herzberg Institute of Astrophysics
Thirty Meter Telescope (TMT)
Space Telescope Science Institute, Hubble Space Telescope (STScI/HST)
National Optical Astronomy Observatory (NOAO)
Gemini North
MIT Lincoln Laboratory
Los Cumbres Observatory Global Telescope Network (LCOGT)
United Kingdom Infrared Telescope (UKIRT)
University of Hawaii, Institute for Astronomy
Advanced Technology Solar Telescope (ATST)
Subaru Telescope
Atacama Large Millimeter/Submillimeter Array (ALMA)
Space Weather Prediction Center (SWPC)
NASA Infrared Telescope Facility (IRTF)

We are very grateful to the individuals from these organizations who provided us with operational insights and best practices.

References

Advanced Technology Solar Telescope (ATST) personnel, telephone and e-mail interviews, June 18, 2013.

Atacama Large Millimeter/Submillimeter Array (ALMA) personnel, telephone and e-mail interviews, September 6, 2013.

Crabtree, Dennis, Herzberg Institute of Astrophysics, telephone and e-mail interviews, July 1, 2013.

Crabtree, Dennis, and Elizabeth Bryson, "Observatory Publications and Citations." Library and Information Services in Astronomy IV (LISA IV), *Emerging and Preserving: Providing Astronomical Information in the Digital Age.* Proceedings of a conference held at Charles University, Prague, Czech Republic, July 2–5, 2002. Edited by Brenda G. Corbin, Elizabeth P. Bryson, and Marek Wolf. Washington, DC: U. S. Naval Observatory, 2003, p. 204–206.

Gemini North personnel, telephone interview, June 25, 2013.

Hawkins, Eric, Director of Engineering, Los Cumbres Observatory Global Telescope Network (LCOGT), telephone and e-mail interviews, June 27, 2013.

Hunsberger, Forrest, Decision Support in Space Situational Awareness, MIT Lincoln Laboratory, telephone and e-mail interviews , June 28, 2013 through July 18, 2013.

Johnson, Carl, OED Division Manager/Archive Project Manager, Space Telescope Science Institute, Hubble Space Telescope (STScI/HST), telephone and e-mail interviews August 21, 2013 through August 23, 2013.

Kerr, Tom, Head of UKIRT Operations, United Kingdom Infrared Telescope (UKIRT), telephone and e-mail interviews, July 18, 2013 through July 19, 2013.

Kwajalein Missile Range personnel, telephone interview, July 18, 2013.

Los Cumbres Observatory Global Telescope Network (LCOGT) personnel, telephone interview June 27, 2013.

MIT Lincoln Laboratory personnel, telephone and e-mail interviews, June 28, 2013 through July 18, 2013.

NASA Infrared Telescope Facility (IRTF) personnel, telephone interview, June 26, 2013.

National Optical Astronomy Observatory (NOAO) personnel, telephone interview, July 18, 2013.

Pacific Defense Solutions, LLC personnel, telephone and e-mail interviews, June 7, 2013, through July 12, 2013.

Panoramic Survey Telescope and Rapid Response System (Pan-STARRS) personnel, telephone and e-mail interviews, June 7, 2013 through July 12, 2013.

Space Telescope Science Institute, "HST Publication Statistics." As of September 17, 2013: http://archive.stsci.edu/hst/bibliography/pubstat.html

Space Telescope Science Institute, Hubble Space Telescope (STScI/HST) personnel, telephone and e-mail interviews, August 21, 2013, through August 23, 2013.

Space Weather Prediction Center (SWPC) personnel, NOAA headquarters, Boulder, Colo., in-person interviews, August 22, 2012 and February 6, 2013.

Subaru Telescope personnel, telephone and e-mail interviews, July 17, 2013 through August 7, 2013.

Thirty Meter Telescope (TMT) personnel, telephone interview, July 19, 2013.

W. M. Keck Observatory, Mainland Observing Notes. As of September 17, 2013: http://www2.keck.hawaii.edu/inst/mainland_observing/

W. M. Keck Observatory personnel, telephone and e-mail interviews, June 3 and June 18, 2013.